ADVANCED MARINE STRUCTURES

ADVANCED MARINE STRUCTURES

Srinivasan Chandrasekaran

CRC Press
Taylor & Francis Group
Boca Raton London New York

CRC Press is an imprint of the
Taylor & Francis Group, an **informa** business

CRC Press
Taylor & Francis Group
6000 Broken Sound Parkway NW, Suite 300
Boca Raton, FL 33487-2742

First issued in hardback 2019

ISBN-13: 978-1-4987-3968-9 (hbk)

Library of Congress Cataloging-in-Publication Data

Chandrasekaran, Srinivas, 1964-
Advanced marine structures / Srinivas Chandrasekaran.
pages cm
Includes bibliographical references and index.
ISBN 978-1-4987-3968-9
1. Offshore structures. I. Title.

TC1665.C456 2016
627'.98--dc23 2015011120

Visit the Taylor & Francis Web site at
http://www.taylorandfrancis.com

and the CRC Press Web site at
http://www.crcpress.com

Dedicated to

My parents, teachers, family members, and friends

Contents

List of Figures

List of Tables

Preface

Marine structures are constructed for different purposes. Recent advancements in construction materials, innovative structural forms, and methods of modeling and analysis make the understanding of the subject more broad-based. This book is an attempt in the direction of explaining various stages of the analysis and design of marine structures. The essential concepts of the subject are presented so as to be useful for practicing engineers and graduate students in different engineering branches, namely: civil, mechanical, structural, naval architecture, production, and marine. With vast experience of teaching this subject and guiding research, a humble attempt is made to present the basics in a closed form that will be useful for readers. Chapters in the book are organized such that the reader gets an overall idea of various types of marine structures constructed for offshore oil and gas exploration, various environmental loads, ultimate load design, fluid–structure interaction, fatigue, and fracture. Numerical examples and application problems are chosen to illustrate the use of experimental, numerical, and analytical studies in the design and development of marine structures. The basics of design, plastic analysis, and estimation of shear center and shape factors will be useful for students because they are explained with simple examples. This book is a repetitive effort in the direction of capacity building for practicing and consulting offshore structural engineers who need to understand the basic concepts of analysis and design of offshore structures through a simple and straightforward approach. Video lectures of the courses are available at: (i) http://nptel.ac.in/courses/114106035; (ii) http://nptel.ac.in/courses/114106036; and (iii) http://nptel.ac.in/courses/114106037, which also complement the classroom mode of understanding of the contents of this book.

Sincere thanks are due to my professors, colleagues, research scholars, my wife, and children, who have given their valuable input, feedback, and moral support to develop the contents of this book. In particular, I wish to express my thanks to Ms. N. Madhavi for her technical assistance and Mrs. Indira for the preparation of the manuscript of the book. I also owe many thanks to all the authors and publishers who have earlier published books on marine structures and allied topics, based on which I developed my concepts on the said subject.

Author

Srinivasan Chandrasekaran is a professor in the Department of Ocean Engineering, Indian Institute of Technology Madras, India. He has teaching, research, and industrial experience of about 24 years, during which he has supervised many sponsored research projects and offshore consultancy assignments both in India and abroad. His active areas of research include the dynamic analysis and design of offshore platforms, the development of geometric forms of compliant offshore structures for ultra-deepwater oil exploration and production, subsea engineering, the rehabilitation and retrofitting of offshore platforms, the structural health monitoring of ocean structures, seismic analysis and design of structures and risk analysis, and reliability studies of offshore and petroleum engineering plants. By invitation of the Ministry of Italian University Research, Dr. Chandrasekaran was a visiting fellow to University of Naples Federico II, Italy, for a period of two years, during which he conducted research on advanced nonlinear modeling and the analysis of structures under different environmental loads with experimental verifications. He has published about 120 research papers in international journals and has refereed conferences organized by professional societies around the world. He has also authored four textbooks: *Seismic Design Aids for Nonlinear Analysis of Reinforced Concrete Structures*, *Analysis and Design of Offshore Structures with Illustrated Examples*, *Advanced Theory on Offshore Plant FEED Engineering*, and *Dynamics of Offshore Structures* (in press), which are popular among graduate students of civil and ocean engineering and are also recommended as reference material for classroom studies as well as research in many universities in India and abroad. Under the auspices of the National Program on Technology Enhancement Learning (NPTEL), Government of India, he has also delivered four web-based courses: Dynamic Analysis of Ocean Structures (http://nptel.ac.in/courses/114106036/), Ocean Structures and Materials (http://nptel.ac.in/courses/114106035/), Advanced Marine Structures (http://nptel.ac.in/courses/114106037/), and Health, Safety and Management in Offshore and Petroleum Engineering (http://nptel.ac.in/courses/114106017/). Dr. Chandrasekaran is a member of many national and international professional bodies and has delivered many invited lectures and keynote addresses to international conferences, workshops, and seminars organized in India and abroad.

About the Book

Marine structures are unique in the field of engineering because they pose challenges in the development and conceptualization of the design. In particular, offshore structures are exceptional because the geometric development of their design forms for oil exploration, production, processing, storage, and transportation are highly innovative. The development of new platforms that initially started onshore has now moved into deeper offshore areas due to the reduced possibilities of new fields in shallower waters. Exploration processes need to become cost-effective, which is possible only by employing innovative design strategies to alleviate the environmental loads that are encountered at greater water depths. Because innovative platform geometries are envisaged, a detailed understanding of their response behavior under various environmental loads is inevitable. The chapters in this book are organized such that the reader gets an overall idea of various types of offshore platforms. Chapter 1 provides details of different structural forms, function, and limitations of offshore platforms including the newly developed FSRUs, offshore triceratops, and environmental loads acting on marine structures. Closed form equations for computing such loads are given for ready reference. Chapter 2 discusses the ultimate load design with illustrations on plastic design, shear center, and theories of failure, leading to a detailed understanding of plastic design. Chapter 3 gives details on fluid–structure interaction, and shows details of numerical models carried out in the research perspective. The hydrodynamic response of offshore structures with perforated members is one of the recent research applications found to be among the effective manners of retrofitting offshore structures. The results of recent research, validated by experimental and numerical studies, are also presented, and innovative ideas that lead to the retrofit and renovation of marine structures are introduced. Chapter 4 is focused on an introduction to reliability as applied to marine structures. Different levels of reliability, along with problem formation, are explained in a convenient style. Chapter 5 deliberates on the application of fracture and fatigue to marine structures. Detailed descriptions of the fundamentals are explained in a classroom style, which is more user-friendly and easier for readers to understand. Basic concepts are emphasized through simple illustrative examples and exercises. Design methodologies and guidelines, which are form-based concepts, are explained through a few applied example structures. One of the interesting approaches the author has adopted is the integration of the concepts with the form-evolved design of offshore structures. Illustrated examples discussed in a classroom style are the outcome of the rich teaching experience the author has in the subject area. The book is also recommended to practicing and consulting offshore structural engineers, who need to revisit basic concepts in a simplified form.

1 Offshore Structures and Environmental Loads

1.1 INTRODUCTION

Many offshore drilling companies are being challenged to check declining production and instead increase production with a new set of discoveries. For some time now, many major operators have seen a reduction in output. To counteract depletion, major oil companies are searching for new reserves in the most challenging regions: deep and ice-infested waters, which are challenging environments for oil exploration (GustoMSC, 2010). Increase in the number of drilling platforms started in 2010 with major orders, which have doubled the demand for deepwater floaters. It is therefore interesting and important to note that only plants with specific capabilities will carry out the quest for deepwater drilling of oil and gas reservoirs. The deepwater regions are located in very remote areas where logistics are another important design consideration. Because the distance between drill sites and ports increases, deck spaces need to be designed with variable load capacities between 10,000 and 20,000 tons. Deck space is very important because more space is required for third-party equipment for well completion and well testing activities, in addition to the space for real drilling equipment. With the new objective to explore and develop more deepwater offshore oil and gas fields and ensure more offshore production, new platforms, such as the DSS series developed by GustoMSC units, are equipped with vertical riser storage (Figure 1.1). The deck space required in such cases is less than that for horizontal storage, thus freeing-up the deck space. Present challenges in the Arctic regions in shallow waters include low temperature, remoteness, ice conditions, hostile ecosystem, and safety.

It is important to note that the development of offshore platforms depends on various factors, namely (i) structural geometry with a stable configuration; (ii) ease of fabrication, installation, and decommission; (iii) low capital expenditure (CAPEX); (iv) early start of production; and (v) high return on investment through increased and uninterrupted production. Offshore plants are unique in many ways. The most important fact is that the newly developed platforms do not have the similar existing ones to compare and understand the design/construction complexities. It is therefore imperative to examine the basics of planning and then select the most suitable structural configuration of the platform. This is one of the essential features of the front end engineering design (FEED).

FIGURE 1.1 Deepwater drilling semisubmersible with vertical riser storage.

1.2 OFFSHORE PLATFORMS: STRUCTURAL ACTION AND FORM IMPROVEMENTS

1.2.1 Bottom-Supported Structures

Energy is the driving force for the progress of civilization. Industrial advancement was first powered by coal and then by oil and gas. Oil and gas are essential commodities in world trade. Oil exploration, which initially started onshore, has now moved to deep waters because of the paucity of resources in shallow waters (Bhattacharyya et al., 2003). To date, there are more than 20,000 offshore platforms of various kinds installed around the world. Geologists and geophysicists search for potential oil reserves beneath the ocean floor, and engineers take the responsibility of transporting the oil from the offshore site to the shore location (Dawson, 1983). There are five main areas of operation from exploration to transportation of oil, namely (i) exploration, (ii) exploration drilling, (iii) development drilling, (iv) production operations, and (v) transportation (Clauss et al., 1992; Clauss and Birk, 1996). Ever since the first offshore structure was constructed, more advanced design technologies have emerged to build larger platforms specifically for deeper waters. Each design is unique to the particular site (Ertas and Eskwaro, 1991). A precise classification of offshore platforms is difficult because of the large variety of parameters involved, such as functional aspects, geometric form, construction and installation methods, and so on. However, platforms are broadly classified based on geometric configurations (Chandrasekaran, 2013a,b,c). Offshore installations are constructed for various purposes, namely (i) exploratory and production drilling, (ii) water or gas injection into the reservoir, (iii) processing oil and gas, (iv) cleaning the by-products of the produced oil for disposal into sea, and (v) accommodation facilities. They are not classified on the basis of their functional use but on the basis of their geometric (structural) form (Sarpkaya and Isaacson, 1981; Sadehi, 1989, 2001, 2007). Because the platforms are designed for greater water depths, their structural form changes significantly; the same form cannot be used at different water depths. This means that the geometric evolution of the platform has to be adapted to accommodate the environmental loads at the chosen water depth (Patel, 1989). Further, the technological complexities faced by new offshore platforms, involving analysis and design, topside details, construction, and installation, are not widely available because they are owned and protected by the respective companies/agencies as part of their copyright. This practice leads to the risk of young practicing engineers, in particular, not being aware of the complexities in designing offshore plants. Hence, prior to the knowledge of FEED it is necessary to understand the different structural forms of offshore structures that have been successful in the past. Because it is well known that each platform is unique in many ways, learning about their structural configurations, limitations with respect to the sea states and water depth, construction complexities, decommissioning issues, and their structural responses is an important stage in pre-FEED (Hsu, 1981; Paik and Anil Kumar, 2007).

The present trend is to design and install offshore platforms in regions that are inaccessible and where the use of existing technologies is difficult (Anagnostopoulos, 1982). The structural form of every platform is largely derived on the basis of structural innovation rather than on the basis of functional advantage. Revisiting existing platforms constructed around the world will impart a decent level of knowledge to offshore engineers (Graff, 1981; Gerwick, 1986). Offshore platforms are classified either as bottom-supported or floating. Bottom-supported platforms can be further classified as fixed or compliant structures; compliant means moving (mobility). Compliancy changes the dynamic behavior of such platforms. Floating structures are classified as neutrally buoyant (e.g., semisubmersibles, FPSO, and monocolumn spars) and positively buoyant (e.g., tension leg platforms). It is important to note that buoyancy plays a very important role in floating offshore structures, because classifications are made on this basis (Bea et al., 1999). Table 1.1 shows the list of jacket platforms constructed worldwide. Fixed platforms are known as template structures, and they consist of the following:

- A jacket or a welded space frame designed to facilitate pile driving and also acting as a lateral bracing for the piles
- Piles permanently anchored to the seabed to resist the lateral and vertical loads that are transferred from the platform
- A superstructure consisting of the deck to support operational activities

The jacket platform complex, shown in Figure 1.2, consists of process, wellhead, and riser equipment, flare support, and living quarters. The relative value of wind,

TABLE 1.1
Offshore Jacket Platforms Constructed Worldwide

Cognac fixed platform	312 m	USA
Pompano	393 m	USA
Bullwinkle	412 m	USA
Canyon station	91 m	USA
Amberjack fixed platform	314 m	USA
Alma fixed platform	67 m	Canada
North triumph fixed platform	76 m	Canada
South venture fixed platform	23 m	Canada
Blacktip	50 m	Australia
CaNgu Vang	56 m	Vietnam
East Belumut A	61 m	Malaysia

wave, and other loads on a typical platform is shown in Figure 1.3. The installation procedures using the launch method of installation are shown in Figures 1.4 through 1.7.

Offshore platforms fall into three major categories, namely (i) fixed platforms, (ii) compliant platforms, and (iii) floating platforms. Different types of fixed platforms are (i) jacket platforms and (ii) gravity platforms. Different types of compliant platforms are (i) guyed towers, (ii) articulated towers, and (iii) tension leg

FIGURE 1.2 Jacket platform complex with process, wellhead, and riser platforms.

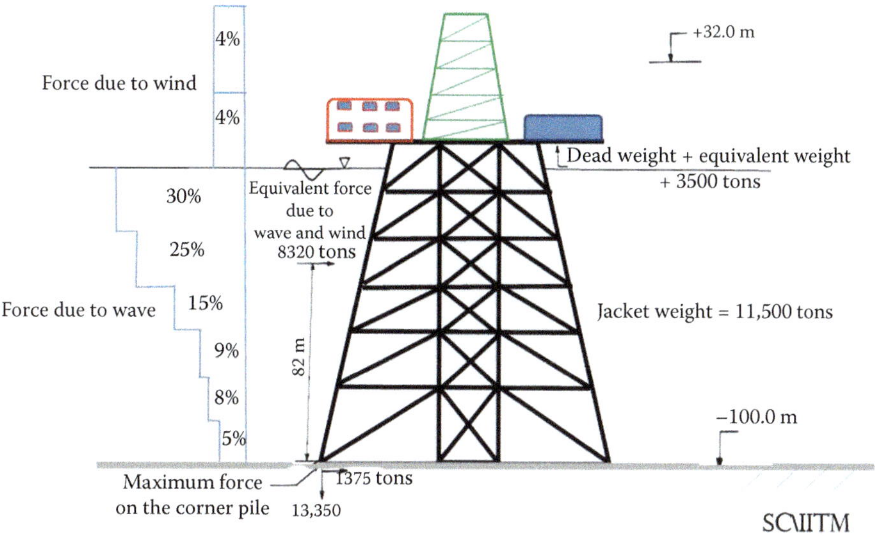

FIGURE 1.3 Typical distribution of wave and wind loads on offshore platforms.

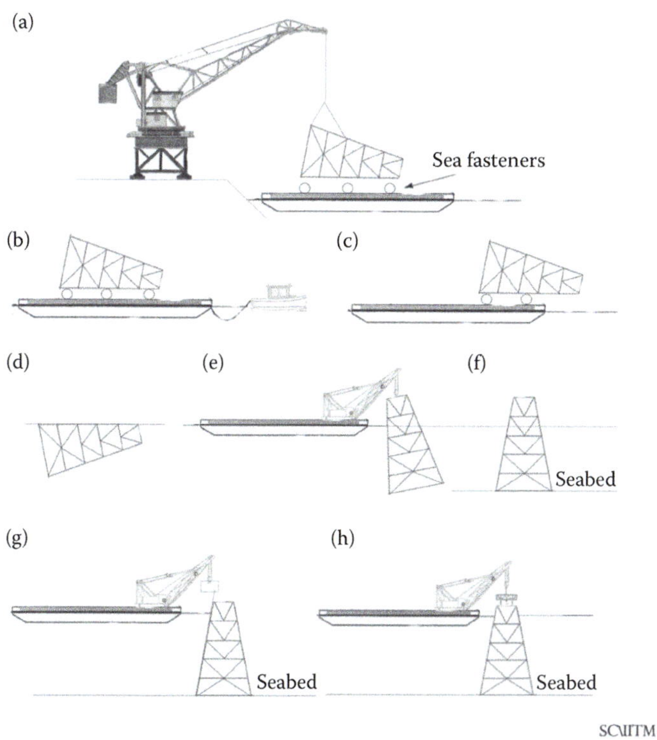

FIGURE 1.4 Installation procedures: (a) loadout, (b) towing, (c) launching, (d) floating, (e) upending, (f) vertical position, (g) piling, and (h) deck mating.

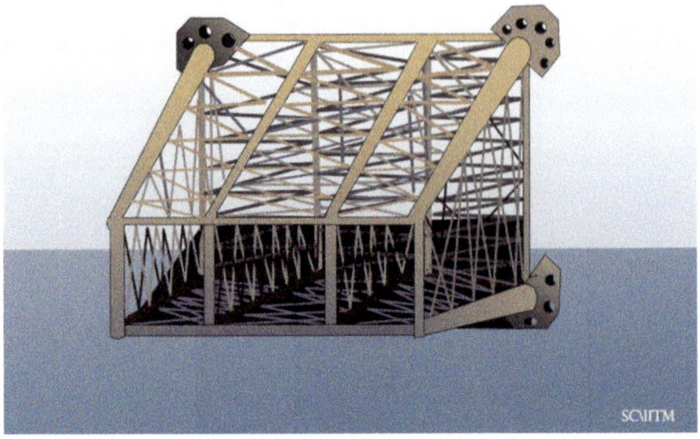

FIGURE 1.5 Transportation of jacket platform.

platforms. Different types of floating platforms are (i) semisubmersibles, (ii) floating production units (FPUs), (iii) floating storage and offloading (FSO) systems, (iv) floating production storage and offloading (FPSO) systems, and (v) spars. Fixed offshore structures are stiffer and tend to attract more forces. Jacket platforms are steel structures and are especially suited for soft-soil regions, for example, clayey soil. They are usually insensitive to the lateral loads by virtue of their rigidity and bottom fixity. Their salient advantages are (i) they support large deck loads; (ii) they have the possibility of being constructed in sections and transported; (iii) they are suitable for large-field and long-term production (can support a large number of wells); and (iv) the piles used for their foundations result in good stability. We can think of these factors as advantageous for a production platform in a remote location in the sea, but they come with a few undesirable characteristics, namely (i) they fail instantaneously without sufficient warning; (ii) they are not suitable for cyclic loading; and (iii) they make the installation process time-consuming and expensive,

FIGURE 1.6 Launching of jacket from a barge after tilting the barge.

FIGURE 1.7 Deck mating by lifting the deck using offshore crane barge.

which delays the start of production. A few other disadvantages are (i) their cost increases exponentially with increase in water depth; (ii) they have high initial and maintenance costs; (iii) they are not reusable; and (iv) the steel structural members are subject to corrosion causing material degradation during service life. Because of their massive geometry, they are also expensive (leading to a high CAPEX). Economic considerations limit the development of fixed (rigid) platforms to water depths of about 500 m. Jacket structures are also known as template structures because the legs act as a template for driving the piles. A few of the important terms associated with this platform are topside or surface facilities, jacket, pipeline, and support services. Topside or surface facilities are the part of the platform that contains the drilling module, the production module, and the crew quarters. Typically the dimensions of the topsides could be 66 m × 66 m per deck level, with four decks with an overall height of 33 m.

1.2.2 Gravity Platforms

In addition to steel jackets, concrete has also been widely used to build some offshore structures. These structures are called gravity platforms or gravity-based structures (GBS). A gravity platform relies on the weight of the structure to resist the encountered loads instead of pilings (API-RP2A, 1989). In regions where driving piles become difficult, structural forms are designed to rely on their own weight to resist the environmental loads. These structures have foundation elements that contribute significantly to the required weight and spread over a large area of the seafloor to prevent failure due to overturning moments caused by lateral loads. Gravity platforms are capable of supporting large topside loads during tow out, which minimizes the hook-up work during installation. Additional large storage spaces for hydrocarbons add to the advantages, which include (i) construction onshore for transport; (ii) towing to the site of installation; (iii) quick installation by flooding; and (iv) use of traditional methods and labor for installation. Table 1.2 shows the list of gravity platforms constructed worldwide. These platforms are also known to be responsible for seabed scouring due to large foundations and for causing severe environmental impact (Chandrasekaran, 2013a).

Gravity platforms have serious limitations, namely (i) unsuitability for sites with poor soil conditions that would lead to significant settlement of the foundation; (ii) long construction periods delaying the start of production; and (iii) natural frequencies falling within the range of significant power of the input wave spectrum (Boaghe et al., 1998). Gravity structures are constructed with reinforced concrete

TABLE 1.2
Gravity Platforms Constructed Worldwide

Name of the Platform	Water Depth (m)
Ekofish 1	70
Beryl A	119
Brent B	140
Frigg CDP1	98
Frigg TP 1	104
Frigg MCP01	94
Brent D	142
Statfjord A	145
Dunlin A	153
Frigg TCP2	103
Ninian	136
Brent C	141
Cormorant	149
Statfjord B	145
Maureen	95.6
Statfjord C	145
Gulfaks A	133.4
Gulfaks B	133.4
Gulfaks C	214
Oseberg	100
Slebner	80
Oseberg North	100
Draugen	280
Heidrun	280
Troll	330

Source: Courtesy of Pennwell Publishing Co.

and consist of a large cellular base surrounding several unbraced columns that extend upward from the base to support the deck and equipment above the water surface (Reddy and Arockiasamy, 1991). Gravity platforms consist of production risers as well as oil supply and discharge lines contained in one of the columns; the corresponding piping system for exchange of water is installed in another column, and drilling takes place through the third column. This particular type is referred to as a CONDEEP (concrete deep water) structure and was first designed and constructed in Norway. The base of the platform is constructed in dry dock after which it is floated and moored in a deep harbor. The construction is then completed by slip forming the large towers in a continuous operation until they are topped off. The structure is then ballasted and a steel prefabricated deck is floated over the structure and attached to its top. The construction of gravity platforms obviously requires deep harbors and deep tow-out channels. The floatation chambers are used as storage tanks, and platform stability is ensured by skirts. Steel gravity platforms exist off Nigeria, where the presence of rock close to the seafloor rules out the possibility of using piles to fix structures to the seabed. Figure 1.8 shows the steel gravity platform located in the Maureen Field, United Kingdom, 2001. The platform is a steel gravity base structure with a weight of 112,000 tons and a height of 241 m, and has steel skirts for penetration into the seabed.

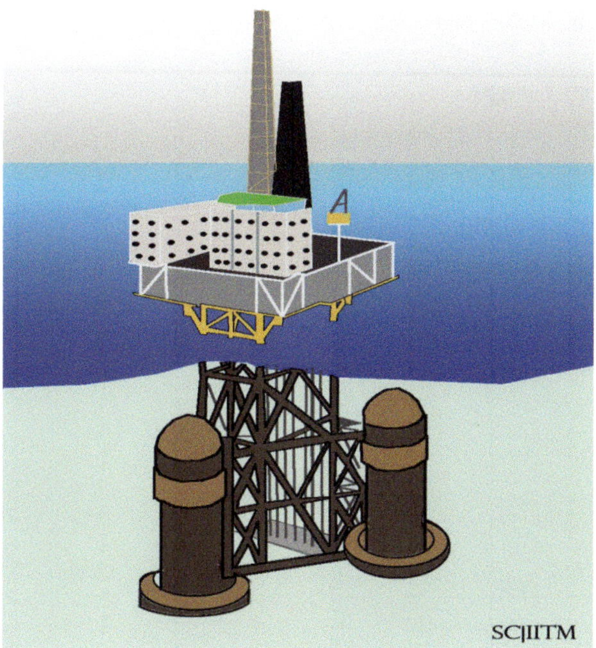

FIGURE 1.8 Maureen Alpha steel gravity platform.

1.2.3 Compliant Platforms

To overcome the negative factors of gravity platforms, we can design a structural form, which will attract fewer forces and remain flexible to withstand cyclic forces. The structural action and the form are modified based on the "mistakes" learned from fixed-type platforms. This is a special kind of reverse engineering, which makes offshore platforms unique. It leads to a continuous improvement from one platform to the next. Hence FEED is on a constant update as new structural forms are being tried for oil and gas exploration in deep and ultra-deep waters (Chandrasekaran, 2013b). Fixed-type offshore structures became increasingly expensive and difficult to install in greater water depths. Hence a modified design concept evolved for structures in water depths beyond 500 m. A compliant tower is similar to that of a traditional platform and extends from the surface to the sea bottom but is transparent to waves. A compliant tower is designed to flex with the forces of waves, wind, and current. The classification of compliant structures includes those structures that extend to the ocean bottom and are anchored directly to the seafloor by piles or guidelines (Mather, 2000). Guyed towers, articulated towers, and tension leg platforms fall into this category. The structural action of compliant platforms is significantly different from that of fixed ones because they resist lateral loads not by their weight but by their relative movement. In fact, instead of resisting the lateral loads, the structural geometry enables the platform to move in line with the wave forces. To facilitate the production operation, they are position-restrained by cables/tethers or guy wires. By attaching the wires to the compliant tower, the majority of the lateral loads will be counteracted by the horizontal component of tension in the cables; the vertical component adds to the weight and improves stability (Dawson, 1983; Chakrabarti, 1994).

1.2.4 Guyed Towers

A schematic view of a guyed tower is shown in Figure 1.9. A guyed tower is a slender structure made up of truss members which rests on the ocean floor and is

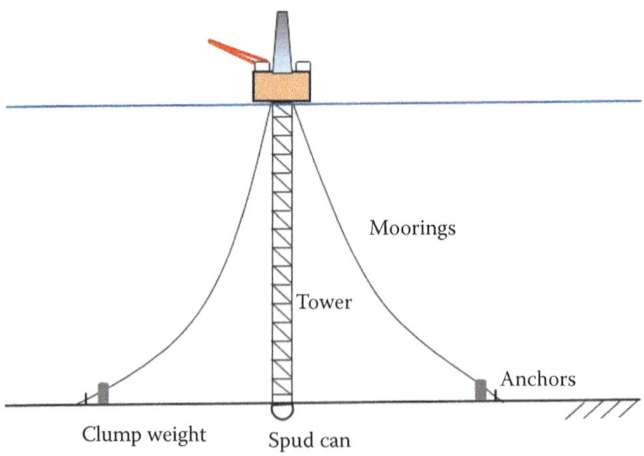

FIGURE 1.9 Schematic view of a guyed tower with its main components.

held in place by a symmetric array of catenary guy lines. The foundation of the tower uses a spud-can arrangement, which is similar to an inverted cone placed under suction. The structural action of the guyed tower makes its innovation more interesting, which is one of the successful form improvements in offshore structural design. The upper part of the guy wirc is a lcad cablc, which acts as a stiff spring in moderate seas. The lower portion is a heavy chain, which is attached with clump weights. Under normal operating conditions, the weights will remain at the bottom, and the tower-deck motion will be nearly insignificant. However, during a severe storm, the weights on the stormward side will lift off the bottom, softening the guying system and permitting the tower and guying system to absorb the large wave loads. Because the guy lines are attached to the tower below the mean water level, close to the center of applied environmental forces, large overturning moments will not be transmitted through the structure to the base. This feature has allowed the towers to be designed with a constant square cross section along their length, reducing the structural steel weight compared with that of conventional platforms (Moe and Verley, 1980).

In 1983 Exxon installed the first guyed tower, named Lena Guyed tower, in the Mississippi Canyon Block at a 300 m water depth. Although the structural form resembles a jacket structure, it is compliant and is moored by catenary anchor lines. The tower has a natural period of 28 s in sway mode while bending and torsion modes have periods of 3.9 s and 5.7 s, respectively. The tower consists of 12 buoyancy tanks of 6 m in diameter and length of about 35 m. A total of 20 guy lines are attached to the tower with clump weights totaling about 180 tons to hold the tower in position. The advantages of guyed towers are (i) lower cost (than steel jacket); (ii) good stability because guy lines and clump weights increase the restoring force; and (iii) possible reuse. The few disadvantages are (i) high maintenance costs; (ii) applicablity to small fields only; (iii) exponential cost increases with increased water depth; and (iv) difficulties in mooring. These factors promoted further innovation in platform geometry resulting in articulated towers.

1.2.5 Articulated Towers

One of the earliest compliant structures, the articulated tower started in relatively shallow waters and slowly moved into deep water. A schematic view of an articulated tower is shown in Figure 1.10. An articulated tower is an upright tower that is hinged at its base with a universal hinge, which enables free rotation about the base. When there was a need to improve the structural form from fixed to compliant, researchers thought of both modes of compliancy, namely (i) rotational and (ii) translational. The capability of large translational motion could make the platform free from being position restrained, and hence rotational compliancy was attempted. In such geometric forms, it is important to note that the design deliberately introduces a single point failure, which is the universal joint (Choi and Lou, 1991). The tower is ballasted near the universal joint and has a large buoyancy tank at the free surface to provide a large restoring force (moment). The tower extends above the free surface and accommodates a deck and a fluid swivel. In deeper water, it is often advantageous to introduce double articulation, the second being at mid-depth (Nagamani and Ganapathy,

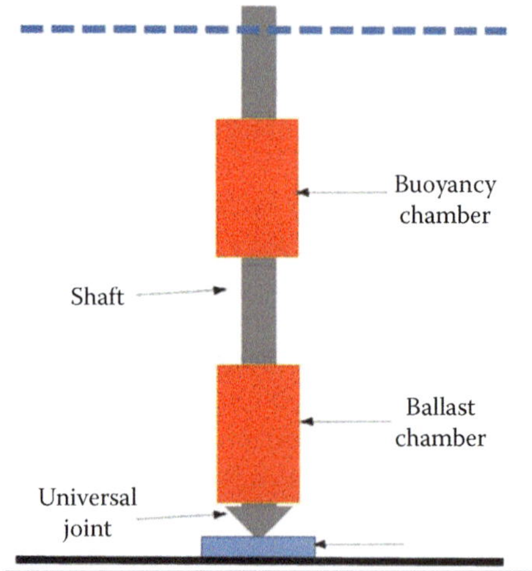

FIGURE 1.10 Schematic view of an articulated tower.

2000). Provision of more articulation reduces the bending moment along the tower (Islam and Ahmed, 2003).

Fatigue is an important criterion for this type of system design because the universal joints are likely to fail under fatigue loads. The advantages of articulated towers are (i) low cost; (ii) large restoring moments due to high center of buoyancy; and (iii) protection of the risers by towers. The few disadvantages are (i) suitability for shallow water only because the tower shows greater oscillations for increased water depth; (ii) inoperability in bad weather; (iii) limitation to small fields; and (iv) universal joint fatigue leading to a single point failure (Chandrasekaran et al., 2010a,b).

In both the above-mentioned structural forms of compliant towers, it can be seen that the structure (tower) extends through the water depth, making it expensive for deep water. Therefore, successive structural forms are intuited toward the basic concept of not extending the tower to the full water depth, but only to retain it near the free surface level as far as possible. In such kinds of structural geometry, it is inevitable that the platform weight is dominant. To improve the installing features and decommissioning procedures, the geometry attempts to be buoyancy dominant instead of weight dominant (buoyancy force exceeds the weight by manifold). While this enables easy fabrication and installation, it also demands skilled labor and high expertise for installation and commissioning of such platforms. The evolved structural geometry is tension leg platforms (Yoneya and Yoshida, 1982; de Boom et al., 1984; Demerbilek, 1990; Vannucci, 1996; Yan et al., 2009).

1.2.6 Tension Leg Platforms

A tension leg platform (TLP) is a vertically moored compliant platform. A schematic diagram of a TLP is given in Figure 1.11. Taut mooring lines, called tendons or tethers, vertically moor the floating platform with its excess buoyancy. The structure is vertically restrained while it is compliant in the horizontal direction, permitting surge,

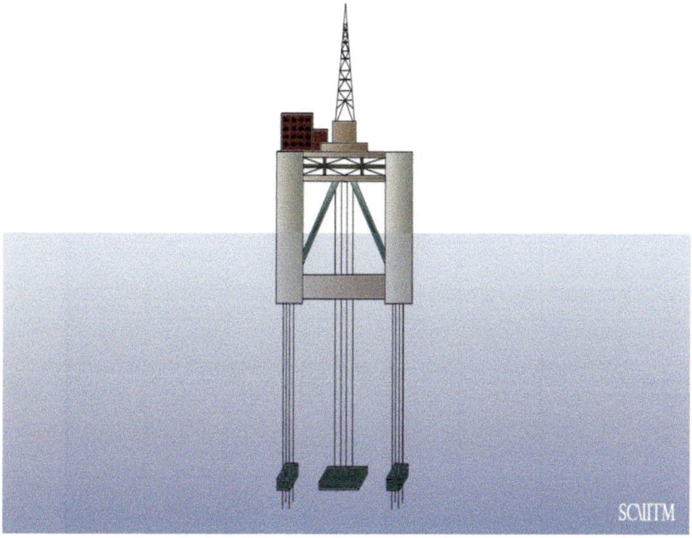

FIGURE 1.11 Typical tension leg platform (TLP).

sway, and yaw motions. The structural action results in low vertical force in rough seas, which is the key design factor (Rijken and Niedzwecki, 1991; Chandrasekaran and Jain, 2002a,b). Substantial pretension is required to prevent the tendons falling slack even in the deepest trough, which is achieved by increasing the free floating draft (Chandrasekaran et al., 2006b). Typical natural periods of the TLP are kept away from the range of wave excitation periods and typically are achieved through proper design (Nordgren, 1987) for TLP resonance periods of 132 s (surge/sway) and 92 s (yaw) as well as 3.1 s (heave) and 3.5 s (pitch/roll). The main challenge for the TLP designer is to keep the natural periods in heave and pitch below the range of significant wave energy, which is achieved by an improved structural form (Kobayashi et al., 1987; Paik and Roesset, 1996; Low, 2009).

Tension leg platform technology preserves many of the operational advantages of a fixed platform while reducing the cost of production in water depths of up to about 1500 m (Iwaski, 1981; Haritos, 1985; Chandrasekaran and Jain 2004; Chandrasekaran et al., 2007a, 2004). Its production and maintenance operations are similar to those of fixed platforms. TLPs are weight sensitive but have limitations in accommodating heavy payloads (Yoshida et al., 1984; Tabeshpour et al., 2006). Usually a TLP is fabricated and towed to an offshore well site where the tendons are already installed on a prepared seabed. Then the TLP is ballasted down so that the tendons may be attached to the TLP at its four corners. The mode of transportation of the TLP allows the deck to be fixed TLP at dockside before the hull is taken offshore (Bar Avi, 1999). Figures 1.12 through 1.16 show a few TLPs that have been built to date.

Under lateral loads like earthquakes, TLPs have shown a favorable response, even though the compliancy in the design is offered in displacement degrees of freedom (Chandrasekaran et al., 2006a; 2007a,b,c; Chandrasekaran and Gaurav, 2008). The

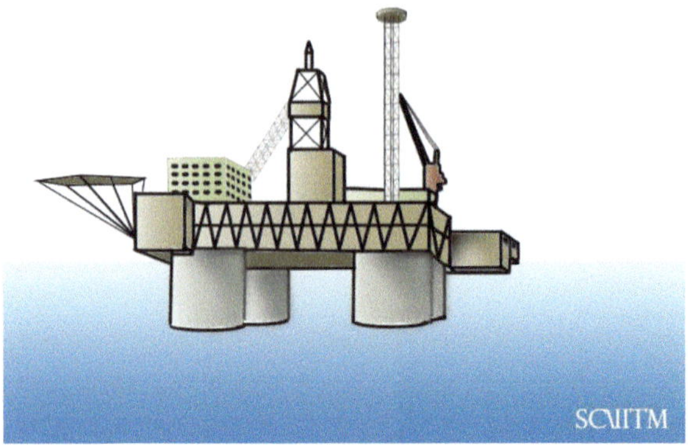

FIGURE 1.12 Heidrun TLP.

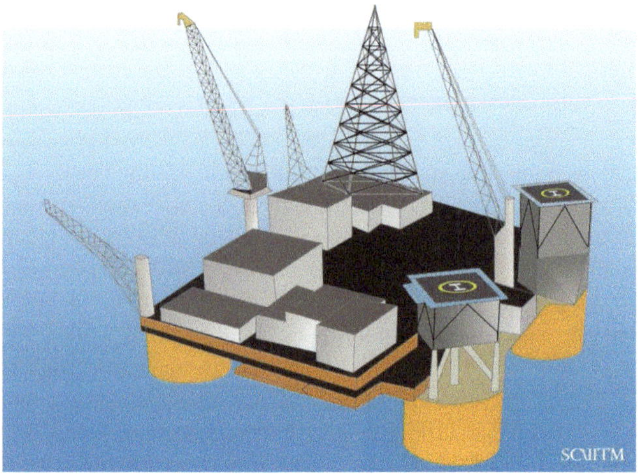

FIGURE 1.13 Auger TLP.

advantages of TLPs are (i) mobility and reusability; (ii) stability, because the platform has minimal vertical motion; (iii) low cost increase with increase in water depth; (iv) deepwater capability; and (v) low maintenance cost. The few disadvantages are (i) high initial cost; (ii) high subsea cost; (iii) fatigue of tension legs; (iv) difficult maintenance of subsea systems; and (v) little or no storage.

1.2.7 Spar Platforms

A spar platform consists of a deep-draft floating caisson, which is a hollow, cylindrical structure similar to a very large buoy. The major components of a spar platform are the hull, moorings, topsides, and risers. Figure 1.17 shows the details of a spar platform. The distinguishing feature of a spar is its deep-draft hull, which produces very favorable motion characteristics. Some salient features of spars

FIGURE 1.14 Snore TLP.

are (i) water-depth capability ranging up to 3000 m; (ii) full drilling and production capabilities; (iii) direct, vertical access production risers (surface production trees); (iv) surface blowout preventer for drilling and workover operations; (v) steel catenary risers (import and export); (vi) inherent stablity because the center of buoyancy is located above the center of gravity; (vii) favorable motion compared with other floating systems; (viii) traditional construction (steel or concrete hull); (ix) cost insensitive to water depth; (xi) potential oil storage; (xii) relocatable over a wide range of water depths; and (xiii) conventional drilling and process components

FIGURE 1.15 Jolliet TLP.

FIGURE 1.16 Hutton TLP.

can be used. The advantages of spar platforms are (i) low heave and pitch motion compared with other platforms (ii) use of dry trees (i.e., on surface); (iii) ease of fabrication; (iv) unconditional stability because the center of gravity is always lower than the center of buoyancy, resulting in a positive GM; and (v) derives no stability from the mooring system and hence does not list or capsize even when completely disconnected from its mooring system. The few disadvantages are (i) difficulty of installation because the hull and the topsides can only be combined offshore after the spar hull is upended; (ii) little storage capacity, which brings along the necessity of a pipeline or an additional FSO; and (iii) lack of no drilling facilities.

1.2.8 Exploratory Platforms

Deepwater offshore platforms are expected to grow from 3% of the total in 2003 to 10% in 2020. Platforms for exploratory drilling have also been improved with advanced structural configurations (Witz et al., 1986). A platform configuration with a semisubmersible is shown in Figure 1.18, and the details of the single buoy mooring are shown in Figure 1.19.

The most commonly used exploration platforms are jack-up platforms, semisubmersibles, and drillships (Figure 1.20). The jack-up platform is used down to a water depth of 90 m whereas semisubmersibles are used down to a depth of 180 m; drillships operate at greater depths, as deep as 2000 m.

The jack-up platform consists of three legs with a deck supporting the helideck, drilling mast (Figure 1.21), and so on. The operation of jack-up platform—drilling a well—is shown in Figure 1.22. The floating deck is used to tow the jack-up platform, with the three legs above the deck, as shown in Figure 1.23. Jack-up platforms are most vulnerable when afloat and under tow during severe weather, whereas they are susceptible to hurricane/cyclone damage while elevated. The punch-through of jack-up rigs is due to sudden failure of the soils while preloading during the operational phase (Issacson, 1982).

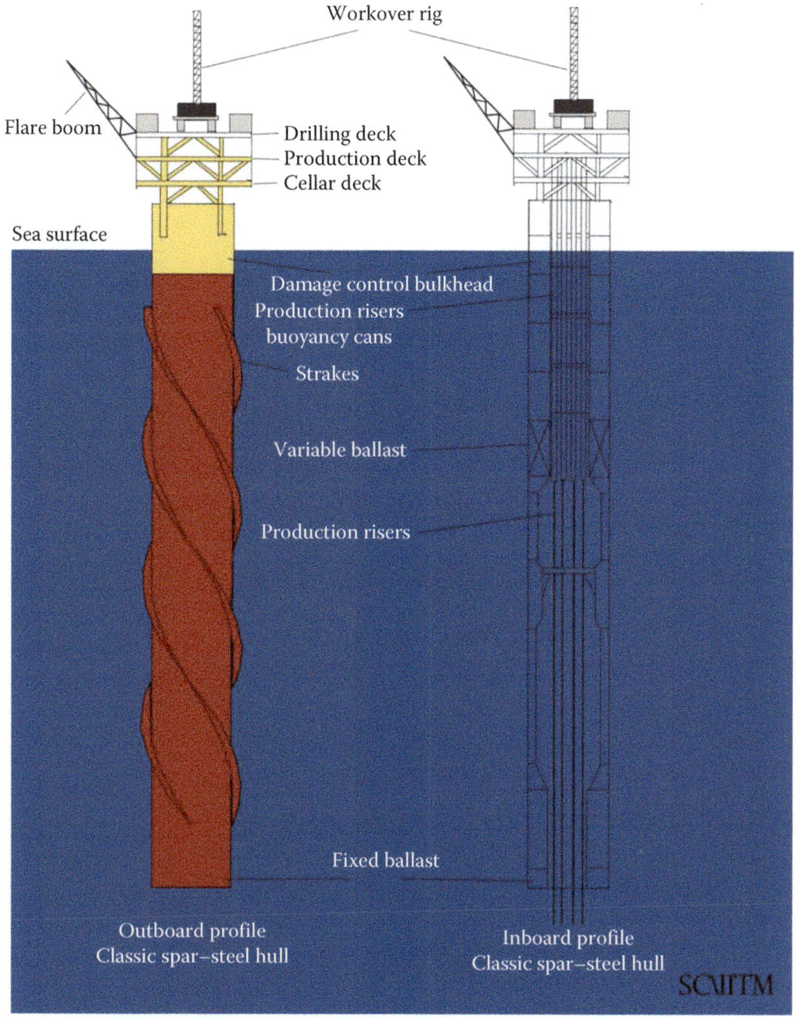

FIGURE 1.17 Conceptual spar.

The buoyancy for a semisubmersible is provided by pontoons, as shown in Figure 1.24, which is kept well below the water surface to reduce the wave action. The semisubmersible is kept in position by mooring lines, as shown in Figure 1.25. Both dry and wet tows can be used for transportation.

The hull is constructed by normal marine and shipyard fabrication methods. In the classic or full cylinder hull form, the whole structure is divided into upper, middle, and lower sections. The upper section is compartmentalized around a flooded center well, which houses different types of risers, namely production riser, drilling riser, and export/import riser. This also provides the required buoyancy while the middle section is configured for oil storage. The bottom section, called the keel, is also compartmentalized to provide buoyancy during transport and to contain any

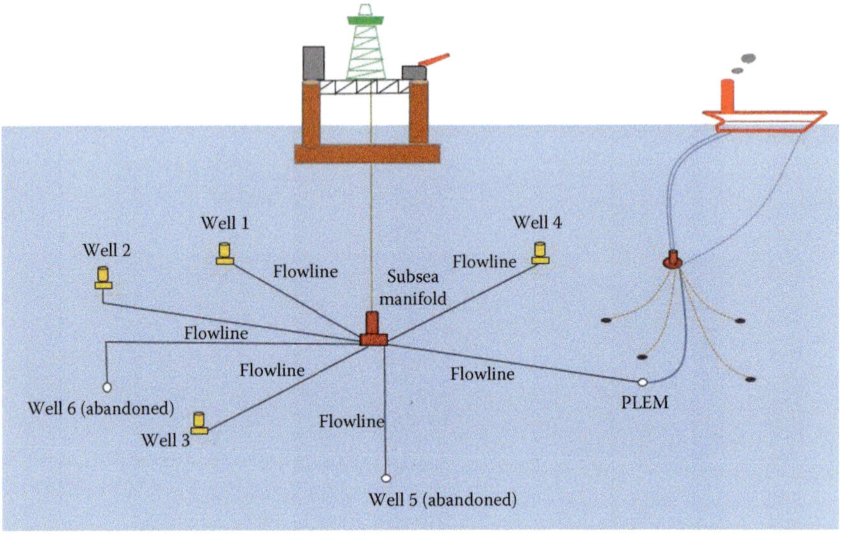

FIGURE 1.18 Semisubmersible, subsea manifold, PLEM, and SBM.

field-installed, fixed ballast. Approximate hull diameter for a typical GOM spar is 130 feet, with an overall height, once deployed, of approximately 700 feet (with 90% of the hull in the water column). The platform position is restrained using a lateral catenary system of 6–20 lines or taut mooring system made possible due to the small motion. Mooring lines are anchored to the seafloor with a driven or suction pile. The footprint created by the mooring system can reach out to a large diameter (typically for a half-mile or more) from the hull to the anchor piles, which can be significantly

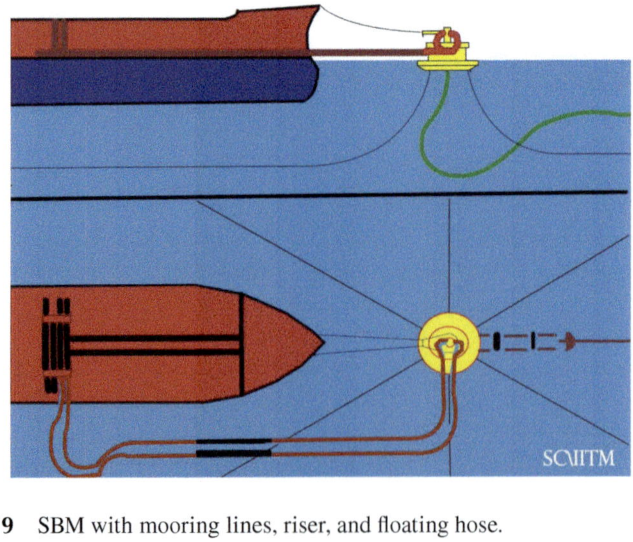

FIGURE 1.19 SBM with mooring lines, riser, and floating hose.

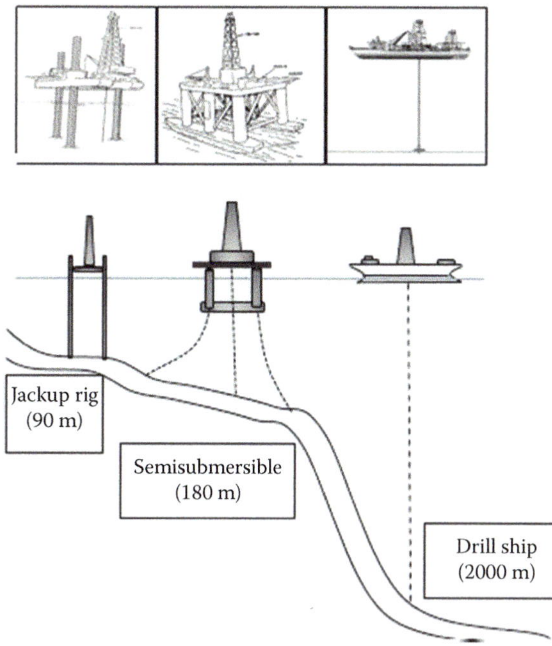

FIGURE 1.20 Jack-up, semisubmersible, and drillship.

reduced by taut mooring lines. This is essentially required to protect the platform from the impact of vessels deployed for oil transportation (Bootan et al., 1987). The topside configuration follows typical fixed platform design practices. The larger topsides consist of drilling, production, processing, and quarters facilities and could also accommodate remote wells/fields being tied back to the spar for processing. The decks can accommodate a full drilling rig (3000 hp) or a workover rig (600–1000 hp)

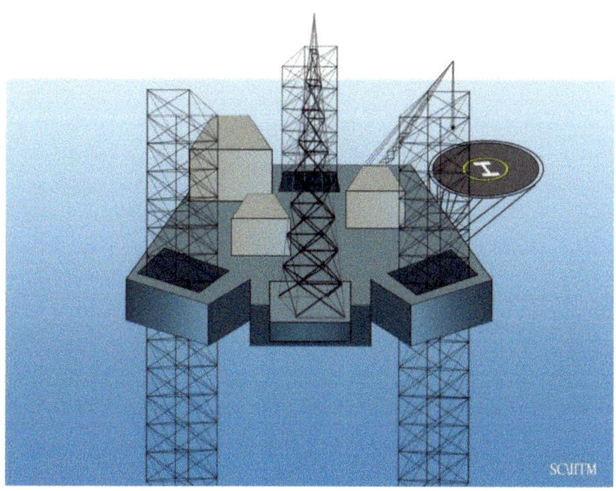

FIGURE 1.21 Jack-up in platform raised position.

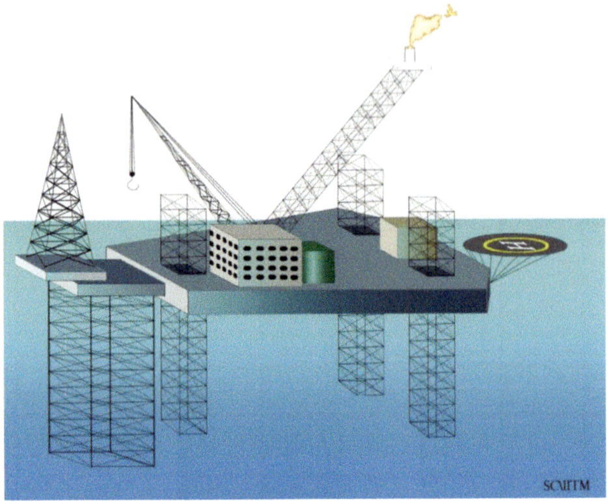

FIGURE 1.22 Jack-up drilling well in a jacket.

plus full production equipment. Total operating deck load—which includes facilities, contained fluids, deck structural and support steel, drilling/workover rig, and workover variable loads—can be about 6600 tons or more.

Floating, production, storage, and offloading systems (FPSO) are typically converted or newly built tankers, which are designed for production and storage

FIGURE 1.23 Jack-up in floating condition with legs raised.

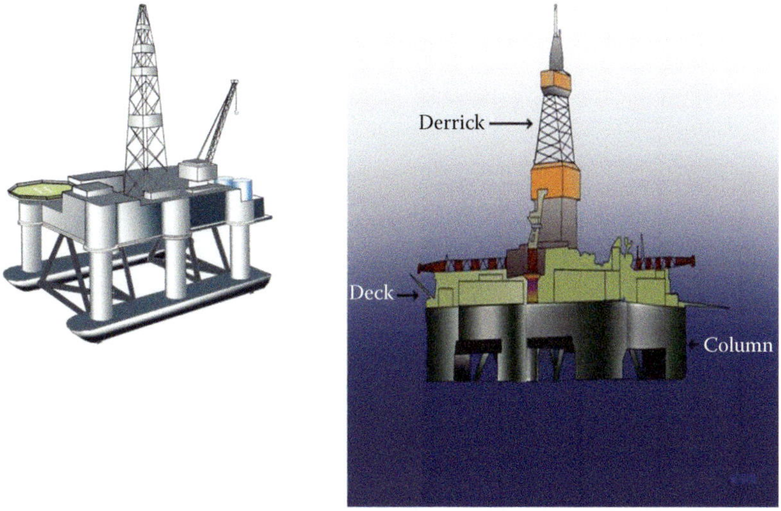

FIGURE 1.24 Semisubmersible with drill rig.

of hydrocarbons. Offloading indicates transfer of produced hydrocarbons from an offshore facility into shuttle tankers or barges for transport to terminals or deepwater ports. An FPSO relies on subsea technology for the production of hydrocarbons, and would typically involve pipeline export of produced gas with a shuttle tanker (offloading) transport of produced liquids. FPSOs are usually ship-shaped structures and are relatively insensitive to water depth. To date, nearly all FPSOs have been installed in water depths less than 1000 m. One recent example was Petrobras' installation of the Marlim Sul Field FPSO in approximately 4700 ft of water located in the Brazilian Campos Basin. Figure 1.26 shows the Marlim Sul Field FPSO deployed by Petrobras. The FPSOs currently in operation cover a wide range of environmental

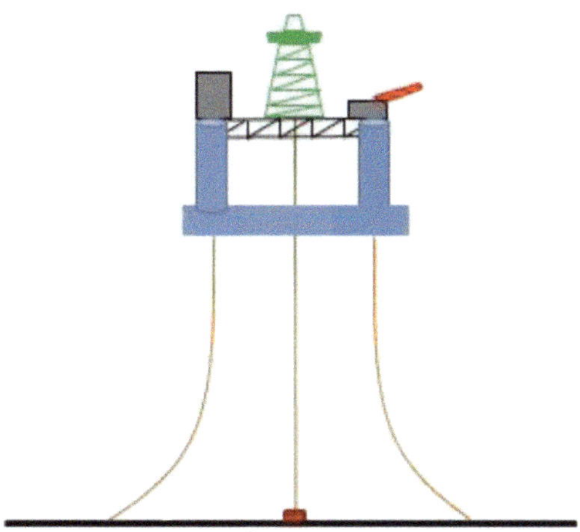

FIGURE 1.25 Semisubmersible with mooring and riser.

FIGURE 1.26 Marlim Sul Field FPSO. (Modified from World Oil.)

conditions, vessel sizes, production rates, operating water depths, and number of risers. The mooring systems of FPSOs are classified as "permanent mooring" or "turret mooring." The majority of FPSOs deployed worldwide are permanently moored, that is, the FPSOs with their moorings and riser systems are capable of withstanding extreme storms in the field location. On the other hand, disconnectable FPSOs have recently attracted more attention. A disconnectable FPSO is typically turret moored. A disconnectable turret is designed for FPSOs to be able to disconnect to avoid certain extreme environments.

The lower part of the turret (lower turret buoy) which connects the mooring lines and risers can be disconnected to allow the FPSO to sail away prior to the approach of a storm. After disconnection, the lower turret with anchoring legs and riser terminations sinks to 50–100 m below the water surface and thus clears off the wave zone. When the extreme weather passes, the FPSO can come back and pick up the submerged mooring/riser buoy and reconnect to it. However, this operation of disconnecting and reconnecting takes considerable time. The primary function of the turret is to allow the FPSO to weather-vane without disrupting the transfer of production fluid between the FPSO and subsea wells. Turret types may be external or internal. FPSOs offer many attractive features, namely (i) environmental hazard is minimized because the platform with crude storage can be disconnected and go to a sheltered area in the event of severe storm; (ii) the lower turret with moorings and riser connections goes below the wave zone and is not significantly impacted by the storm; (iii) the mooring system and its associated turret components can be of smaller size because it only needs to be designed against less severe storms when the FPSO stays connected; and (iv) greater safeguards against platform damage and liability to the operator in the event of a storm. Advantages are (i) low cost; (ii) mobility and reusability; (iii) reduced lead time; (iv) quick disconnecting capability which can be useful in iceberg-prone areas; (v) little infrastructure required; and (vi) turret mooring system enables FPS (converted ship type) to head into the wind/waves reducing their effect. The few disadvantages are (i) limited to small fields; (ii) low deck load capacity; (iii) damage to risers due to motion; (iv) poor stability in rough seas; and (v) little oil storage capabilities. Deepwater platforms have processing equipment to facilitate production equipment. Storage of oil is the main concern in offshore operations. Once the whole drilling operation is complete, the drilling platform becomes a well-protector and a storage platform. Mostly, a tanker ship is anchored next to the treatment platform and serves as oil storage. Personnel employed in the offshore platforms are generally transported by boats or helicopters; hence all platforms must be provided with helipads, mooring bits, bumpers, cranes, stairs, and so on, for use with workboats and crew boats. Crew transported by helicopter are generally spared seasickness on arrival and remain prepared for work. Apart from considerable time savings and reduction in cost, helicopter reliability and capability in bad weather is much better than that of boats. Offshore platforms should be generally designed with a large helicopter landing area to handle loading and unloading operations easily and quickly. Basic dimensions of the helipad on the platform deck depends on the overall length of the helicopter; usually a circular pad of diameter equal to the overall length of the helicopter is considered for the design. The heliport landing area must be large enough to handle loading and unloading comfortably. Provision is generally made on the deck to enable a slant take-off of helicopters. A typical offshore helicopter is some 24 m × 24 m to 49 m × 49 m in size. The offshore deck should be designed for a helicopter load in one of two ways. Either it is designed for a concentrated load up to 75% of the gross weight of the largest helicopter or for an impact factor twice the weight of the largest helicopter. Wind indicators are provided for the heliport to indicate the true wind direction, along with the landing boundaries marked in yellow. Helipads are preferably located at one corner of the platform to facilitate easy landing and takeoff. It preferable to locate the power units, flare stack, and drilling units/equipment far away from the location of helipad to reduce possible risk during helicopter operations.

1.3 NEW-GENERATION OFFSHORE PLATFORMS

Because of nonavailability of oil and gas reserves in medium waters depths, oil and gas exploration is targeted at deep waters and ultra-deep waters. Because the encountered environmental loads are more severe in greater water depths, the geometric form of offshore platforms proposed for deep waters and ultra-deep waters needs special attention. Apart from being cost-effective, the proposed geometric form shall also have better motion characteristics under the encountered forces arising from rough seas. Offshore structures that are found to be suitable for deep water and ultra-deep water are shown in Figure 1.27.

1.3.1 Buoyant Leg Structures

Buoyant leg structures (BLSs) are tethered spars with single or groups of cylindrical water piercing hulls; these are structural forms alternative to TLPs and conventional spars. They are positively buoyant, with the buoyancy exceeding the mass of the structure. Although positively buoyant, positive metacentric height is maintained

FIGURE 1.27 Different types of ultra-deepwater structures.

to ensure the desired structural stability even after the removal of tethers from the structure. This characteristic ensures high stability and deep draft, which makes the structural form relatively insensitive to increased water depth. Because the BLS is a deep-draft structure, the exposed structural part near the free surface is reduced as are the forces exerted on the structure when compared to conventional TLPs. Because the risers are inside the moon pool of the BLS, the forces exerted on the risers are also minimized, but below the keel of the BLS wave and current forces act. Halkyard et al. (1991) initially proposed the concept of a tension buoyant tower, shown in Figure 1.28, which was subsequently modified by other researchers (Robert and Cuneyt, 1995; Perryman et al., 1995). The structural form of BLSs has evolved by combining the advantageous features of spars and TLPs where the deep-draft hull limits the vertical motion to a significant extent (Shaver et al., 2001); a BLS resembles a spar

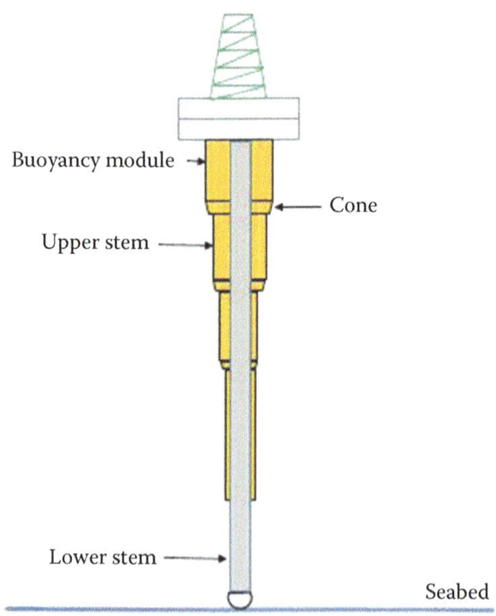

FIGURE 1.28 Tension buoyant tower.

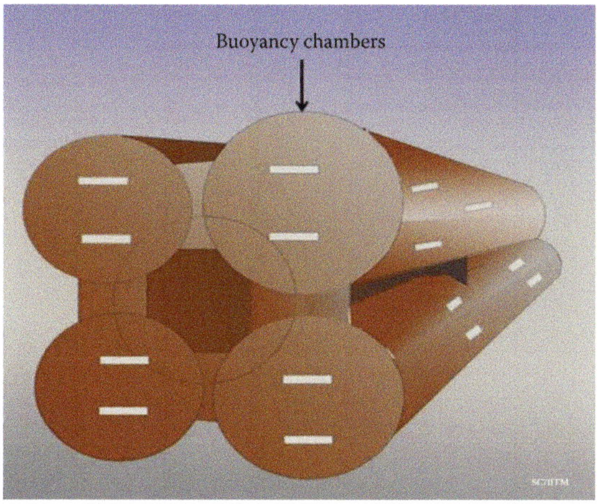

FIGURE 1.29 Buoyant tower in the fabrication yard.

due to its shape and deep-draft feature, and its response behavior is similar to that of a TLP due to its restoring system. BLSs are simple to fabricate, easy to loadout, tow, and install (Capanoglu et al., 2002). Figure 1.29 shows views of the buoyant tower in the fabrication yard, while different stages of installation of BLS are shown in Figure 1.30. The installation process of BLSs is a combination of the installation procedures of spars and TLPs. Because a spar is a stable structure, it is installed simply by free floating while a TLP is generally installed to achieve required pretension in the tethers using (i) ballast; (ii) pull-down; or (iii) both pull-down and ballast methods. During the installation of BLSs the structure can be free floated using its permanent ballast. Pretension in the tethers can be achieved by the above-mentioned procedure. In the ballast method, the structure will be additionally ballasted until it achieves the required draft; the tethers are then attached from the structure to the seafloor. Additional ballast will be removed from the structure to enable pretension in the tethers. In the pull-down method, the free floating structure will be pulled down until it achieves the required draft; excess buoyancy that is transferred to the tethers enables the desired pretension. The pull-down and ballast method is a combination of the two above procedures. BLS gives improved motion characteristics and more convenient riser systems, because they consist of simple hulls in comparison to spars or TLPs. BLSs are more economic than TLPs or spars due to the reduced cost of commissioning. The first buoyant tower drilling production platform, CX-15 for Peru's Corvina offshore field was installed in September 2012 at a water depth of more than 250 m with a production capacity of 12,200 barrels per day.

1.3.2 Triceratops

More innovative geometric forms of offshore platforms have evolved in the recent past to improve the motion characteristics of these platforms in deep water and ultra-deep water; triceratops, nonship-shaped FPSOs, and Min Doc are just a few of them. The conceptual idea of one triceratops, discussed in the literature

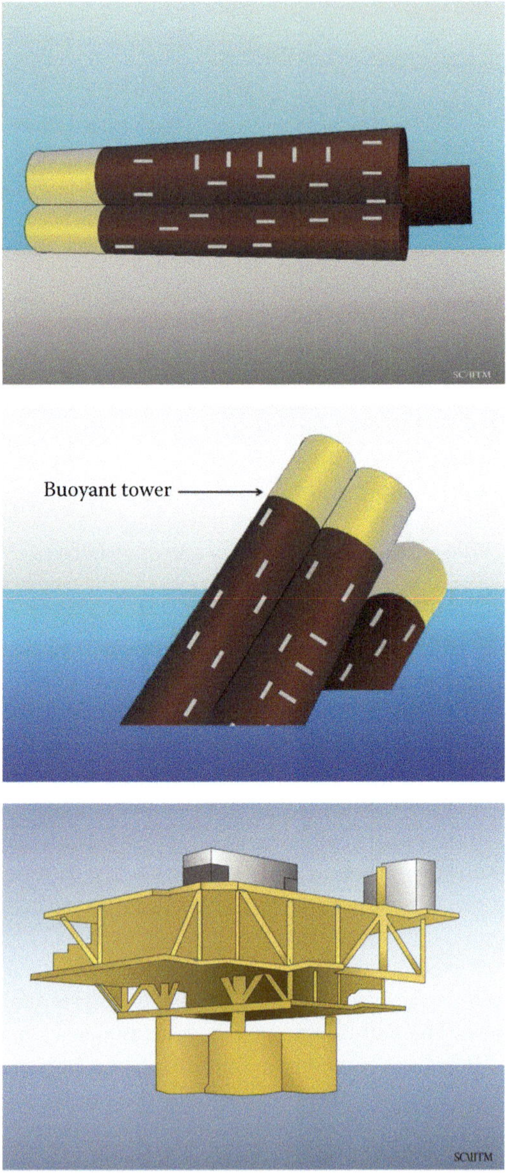

FIGURE 1.30 The first buoyant tower CX-15 loadout and installed structure in Peru's Corvina offshore field.

indicates favorable characteristics of the platform in deep water and ultra-deep water (White et al., 2005); Figure 1.31 shows the conceptual view of triceratops. Geometric innovativeness, imposed on the design by the introduction of ball joints between the deck and the BLS, makes triceratops different from other new-generation offshore platforms.

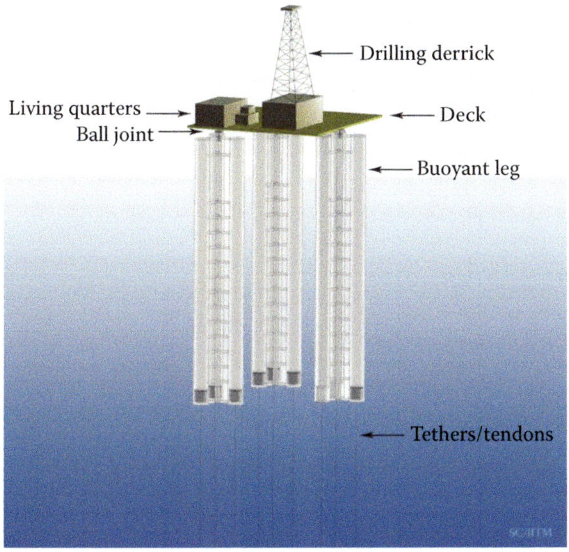

FIGURE 1.31 Conceptual view of triceratops.

Triceratops consists of three BLS units, a deck, three ball joints between the BLS units and the deck, and a restoring system either with restraining legs or with tethers. Ball joints transfer all translation but not rotation about any axis, making the platform different from other classic types of offshore structures. The different motion characteristics of structural members, such as BLS and deck, give uniqueness to its structural behavior under lateral loads.

Common types of offshore platforms have rigid body motion due to the rigid connection between the members, which makes the platform respond as a single unit. Because triceratops has different integrated structural elements, it behaves as a rigid body in all translations but not in rotations about any axis due to the presence of the ball joints; the rotational responses of the BLSs differ from that of the deck. Studies focusing on their response behavior become interesting because the responses of BLS and deck have to be dealt with separately, which is not common practice in most offshore platforms. In addition, the derived geometric form imposes a few advantages namely (i) reduction in forces exerted on the platform due to the decrease in the exposed part of the structure near the free surface and (ii) protection of risers from lateral forces because they are located inside the moon pool of the BLS. The presence of ball joints between the deck and BLS restrains the transfer of the rotational motion of the deck from BLS; the translation and rotation motion of the BLSs under the encountered environmental loads are significantly high. However, due to the deep draft of the BLSs, there is more possibility of unusual corrosion. Corrosion challenges can be dispatched by a few techniques, namely (i) frequent inspection using corrosion testing probes; (ii) use of sacrificial anodes; (iii) anticorrosive coatings; and (iv) use of cathodic protection. The salient advantages are namely (i) better motion characteristics; (ii) suitability for deep waters; (iii) improved dynamics in comparison to TLPs and spars; (iv) wells are within protected environments and are laterally supported; (v) simple structure; (vi) simple station keeping; (vii) easy installation and decommission (installation can be part by part or as a whole structure); (viii) reusable and relocatable; (ix) simple restraining system (does not require high-strength systems like TLPs); (x) highly stable structure, and (xii) relatively low cost.

1.3.3 Floating, Storage, and Regasification Unit

Transportation of unprocessed crude oil from the drilling/exploratory platform to the onshore site involves expensive systems such as pipes, large vessels, and so on, which make oil production more expensive. In particular, they prove to be highly uneconomical for the offshore platforms located far offshore. The key components of floating, storage, and regasification units (FSRUs) consist of regasification equipment that transform a LNG at –160°C to that at high pressure, storage tanks, loading arms for receiving LNG, export manifolds, and seawater pumps that use seawater to regasify the LNG. FSRUs are the more cost-effective alternative to meet the low demand of LNG than traditional, land-based terminals. They contain a regasification unit, a gas turbine with generator, air compressors, fuel pumps, fire water and foam systems, freshwater systems, cranes, lubrication oil system, lifeboats, and a helipad. The LNG is stored at –160° in double-walled insulated tanks to limit boiloff. The outer walls of the tank are made of prestressed reinforced concrete or steel to limit the temperature effect during the storage period. Despite the high-quality insulation, a small amount of heat still penetrates the LNG tanks, causing minor evaporation. The resulting boil off gas is captured and fed back into the LNG using a compressor and recondensing systems. This recycling process prevents any natural gas from escaping the terminal under normal operating conditions. The LNG is subsequently extracted from the tanks, pressurized, and regasified using heat exchangers. The tanks are equipped with submerged pumps that transfer the LNG to other high-pressure pumps. The compressed LNG (at around 80 times atmospheric pressure) is then turned back into a gaseous state in vaporizers. Once returned to its gaseous state, the natural gas is treated in a number of ways, including metering and odorizing, before it is fed into the transmission network.

The LNG is simply warmed using the heat from seawater. This process is carried out in a heat exchanger (with no contact between the gas and the seawater) producing only a slight drop in the temperature of the seawater, which reaches 6°C at the end of the discharge pipe, quickly becoming imperceptible once diluted. Natural gas is odorless. Although nontoxic, it is inflammable and is therefore odorized to ensure even the slightest leak can be identified. This is done at the terminal, before the natural gas is distributed, by injecting tetrahydrothiophene (THT), which is an odorant detectable in very small doses.

Gas turbine equipment on the topside of the FSRU uses multiple units with generating capacity of up to 10–12 MW. The instrument air system provides air for the plant and instrument air for process control and maintenance. Inert gas (nitrogen) is generated on demand by a membrane package using dry, compressed air. A backup inert gas supply system consisting of compressor seals, cooling medium, expansion drums, and utility stations is also provided. The oil pump provides high pressure oil to the engine. The fuel is pumped from the fuel tank to the primary fuel filter/water separator, which is then pressurized to 650 kPa gauge pressure by the fuel transfer pump. The pressurized fuel is then sent through the secondary/tertiary fuel filter. The water supply for the fire-fighting systems is supplied by fire water pumps at a pumping rate of about 600–5000 m^3/h at the discharge flange at a pressure of about 18 bar. A film-forming fluoroprotein (FFFP) concentrate system is provided to enhance the effectiveness of the deluge water spray that protects the separator module, which has a high potential for hydrocarbon pool fires. FFFP is a natural protein foaming agent that is biodegradable and nontoxic. The freshwater maker system utilizes a reverse osmosis process to desalinate seawater at the rate of 5 m^3 per hour. The saline effluent from the freshwater is directed overboard through the seawater discharge caissons, while the freshwater is stored in a freshwater tank. Water delivered to the accommodation module is further sterilized in a UV sterilization plant before being stored in a potable water header tank. The lube system contains an oil cooler, oil filter, gear-driven oil pump, pre-lube pump, and an oil pan that meets offshore tilt requirements. An internal lubrication system is designed to provide a constant supply of filtered, high-pressure oil. The system meets the tilt requirements for nonemergency offshore operations. Lubrication oil should have special features in offshore requirements, namely (i) water solubility; (ii) non-shearing on water surface; (iii) excellent lubrication properties; (iv) biodegradability; and (v) nontoxicity to aquatic environment.

1.4 CONSTRUCTION CHALLENGES IN OFFSHORE PROJECTS

The ocean presents a unique set of environmental conditions, which dominate the methods, equipment, support systems, and construction methods to be employed for the construction and installation of offshore platforms. Unfortunately, relatively small attention is paid to the influences of environmental factors on the construction of offshore platforms. Because the construction projects of offshore structures are subjected to the critical combination of many environmental forces continuously during the erection and commissioning stages, it is important to understand the effects of a combination of forces on the structural members, particularly during the construction. Various factors that influence the construction methods of offshore projects are (i) distance from the shore (offshore distance); (ii) water depth at which they are installed; (iii) hydrostatic pressure and buoyancy forces; (iv) temperature; (v) seawater and air interface chemistry; (vi) currents, (vii) waves and swells; (viii) wind and storm; (ix) rain, snow, whiteout, atmospheric icing, lightening; (x) ice and icebergs; and (xi) seismicity of the installation site.

Geotechnical aspects also dominate the choice of foundation system and method of construction. In many offshore foundation sites, the in-place strength will be quite different (found greater) from the ones that estimated by conventional methods. Thanks to advanced methods, such as the electric receptivity measurements, sampling methods to obtain the seafloor characteristics have improved.

Construction engineers need to recognize these problems so they can adequately interpret the geotechnical reports and logs of boreholes, which helps them make appropriate decisions regarding construction methods, selection of equipment, and so on. Failure to recognize the potential problems has led to a number of serious cases of cost overruns and delays in construction schedules. Layout of offshore construction projects inherently inculcates another issue related to significant variations in the geotechnical properties at the construction site. This is due to the fact that the platform layout extends over a substantial area, which results in variations of the geotechnical properties of the seafloor. Because of the lack of time and the high cost involved, offshore construction engineers can place undue emphasis on only a few borings. Instead, it is preferable to use advanced methods such as sparker surveys to study the variations of site geology over a larger area. The nature of the soil profile in the seafloor, as detailed from the bore hole, is (i) dense sand; (ii) calcareous sand; (iii) boulders near the sea floor; (iv) overconsolidated silts; (v) subsea permafrost and clathrates; (vi) weak arctic silts and clays; (vii) ice scour and pingos; (viii) methane gas; and (ix) muds and clays. The quality of muds and clays poses special kinds of problems in the construction projects. A few of the technical issues are (i) deciding the underwater slopes; (ii) pile driving depth; (iii) short-term bearing strength; (iv) dredging, sampling, and penetration; and (v) consolidation of clay, leading to the improvement in scour strength.

Offshore construction and installation are subjected to serious constraints from nature in terms of protection of the built environment. Offshore construction projects are carried in an environment, which teems with the marine life. In the past, many projects were subjected to serious criticism for not considering the factors of marine life interface in advance; the approach has been one of amelioration of the problem rather than integration. Not giving due consideration at the planning stage in advance attracts a lot of restrictions and penalties, which delay construction significantly.

1.4.1 Offshore Construction Equipment

A variety of modern construction methods and equipment make offshore construction challenging and novel. The availability of construction equipment of higher capacity has, in turn, played a major role in altering the methods and systems employed. Safety being the primary concern in offshore construction, engineers must understand the capabilities and limitations of the equipment that is deployed. They should remain alert to detect the early signs of problems before they grow to catastrophic proportions; a full understanding of the performance of the equipment is mandatory to avoid accidents in offshore projects. Choice of construction equipment for offshore construction depends on a number of critical factors, namely (i) motion response; (ii) buoyancy; (iii) draft free-board; (iv) stability; (v) precision; and (vi) damage control. The list of equipment commonly used in construction projects is (i) barges; (ii) dredging apparatus; (iii) boats; (iv) drilling vessels, and (v) special purpose equipment. Barges are of various types, such as crane barges, offshore derrick barges, semisubmersible barges, jack-up construction barges, launch barges, and pipe laying barges. Boats are also of a number of types, such as supply boats, crew boats, anchor-handling boats, and tow boats. Figure 1.32 shows a semisubmersible

FIGURE 1.32 Semisubmersible crane barge.

barge with a capacity of 10,000 tons with tandem lift. Figure 1.33 shows a floating crane barge used in construction projects. A floating crane barge is a ship fitted with specialized equipment for loading–unloading, trans-shipping, and transportation of heavy loads. The larger floating crane barges are mostly semisubmersibles, but conventional monohulls are also used.

FIGURE 1.33 Floating crane barge.

FIGURE 1.34 Offshore mast crane AHC.

Figure 1.34 shows a special mast crane of 5000 MT capacity with a large turning radius. The use of larger capacity barges for pipe laying is a common practice in offshore construction projects. Figure 1.35 shows a typical pipe laying barge carrying a large stack of pipes to be laid in a major construction project. All the necessary equipment required for pipe laying, including stocking them in bulk, are inherent features of such large barges.

FIGURE 1.35 Pipe laying barge.

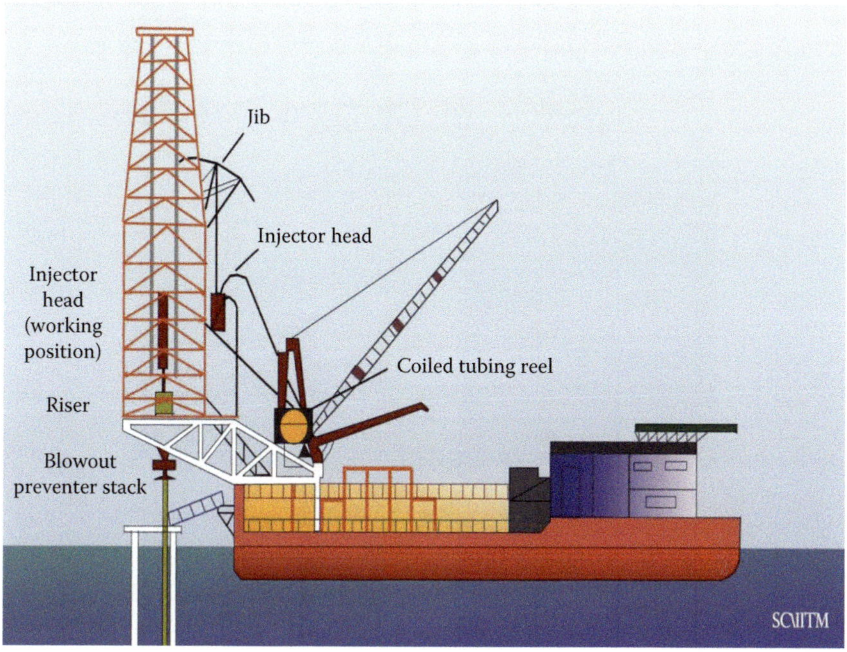

FIGURE 1.36 Typical drill barge.

Drill barges are also commonly seen in offshore installation. Figure 1.36 shows the components of a drill barge. Offshore lifting barges are used to install multiple modules on the topside of the platform (Figure 1.37). Large barges are deployed to transport topside equipment to the plant (Figure 1.38).

The basic set of operations, common to all offshore construction projects, are towing, mooring and anchoring, ballasting, handling heavy loads at sea, transferring

FIGURE 1.37 Lifting barge used in installation of modules of topside.

FIGURE 1.38 Large barges used for loading/transporting topside equipment.

personnel to the site at sea, offshore surveying and navigation, diving and underwater works systems, underwater concreting and grouting, and temporary buoyancy augmentation during construction.

1.5 FOUNDATION SYSTEMS AND SEABED ANCHORS

Offshore foundations are different from conventional foundations because the seabed needs to be investigated for its capacity to withstand the weight of the structures and the applied loads. Most of the early types of offshore structures are bottom-supported. They are held in position in the seabed either by pile foundation or by their own self-weight (GBS platforms). Bottom-supported structures use piles or their own weight to resist wave loads (lateral loads). The latter case can result in serious seabed scouring and settlement of foundations, which are special problems related to offshore geotechnical engineering. The offshore environment poses many challenges to geotechnical engineers namely (i) expensive site investigation; (ii) the presence of carbonates make the seabed conditions complex for foundation systems; (iii) the capacity to resist large lateral loads arising from waves and current; and (iv) less freedom of design modifications, because they may result in expensive construction schemes. One of the serious problems faced in the foundation of offshore plants is the nature of the soil. Usually (and mostly), soil is made of sediment deposits whose layered thickness varies in different regions. Their compaction and consolidation strength are difficult to determine, and therefore foundation design becomes complex. In addition to the effects of waves on offshore structures, currents are responsible for eddies around a structure, which can cause scouring and erosion of the soil. Geohazards are associated with geological activity, geotechnical features, and environmental conditions (Hitchings et al., 1976; Hoeg, 1976). Information on their potential risks is acquired through the study of geomorphology, the geological setting, and the tectonic framework. In addition, geophysical and geotechnical surveys of the seafloor are also carried out for more clarification. Examples of potential threats include tsunamis, landslides, active faults, and soil layering. Soil investigation is a vital and essential part of the construction of offshore structures. Because soil ultimately resists the forces and movements present in the piling at the bottom of the ocean, soil investigation plays a vital part in the design. The under seabed soil will normally be clay, sand, silt, or a mixture of these. Soil is obtained from borings at the site and subsequent tests are carried out on samples. Every project must acquire a site-specific soil report containing soil stratification and its characteristics for load bearing in tension and compression, its shear resistance, and the load–deflection characteristics of axially and laterally loaded piles. The soil report should show the calculated minimum axial capacities for piles of the same diameter as the platform design piles, SRD (soil resistance to driving) curves, different types of mud-mat bearing capacity, pile group action curves, shear resistance values, and pile tip end bearing values. Pile axial capacity values are normally called "T-Z" values, shear values are called "P-Y" values, and end bearing values are called "Q-Z" values. Engineers use these values as input into the structural analysis model (normally in StruCad, FASTRUDL, or SACS software) and hence will determine minimum pile penetrations and size considering a factor of safety of 1.5 increased to 2.0 for operating loads. The unity check ratios must be within 1.0 in the piles or anywhere else in the platform as per API, 1996 (Sadehi, 2007).

Foundation design is critical for bottom-supported structures like GBS platforms due to their large area of contact. One of the primary requirements of a good foundation system is nonlifting. This can only be ensured if the stabilizing moment from the large weight of the platform increases continuously with respect to the overturning moments arising from the lateral loads. In GSB, this is taken care of by large caissons, which are under permanent ballast conditions either by seawater or stored oil. During storm magnitude waves, bottom-supported platforms resting on clay or silt deposits show cyclic horizontal displacements and rotations about their base. This will result in a gradual increase in the vertical settlement as well. For foundations resting on clay layers, weak slipping strength is counteracted by driving skirt piles. Skirts serve three purposes, namely (i) they penetrate the weaker section of the soil along the depth of foundation; (ii) they transfer the lateral load to the stronger soil beneath; and (iii) they protect the foundation from scouring (Chandrasekaran and Pannerselvam, 2009). Of all the failure modes of offshore foundations, sliding is the simplest. Skirts that penetrate deeper into the seabed can prevent sliding to a greater extent. Other failure modes of offshore foundation are bearing capacity failure, rocking, and liquefaction. When the shear strength of the soil is weak, as in the case of clay soil, failure in the bearing capacity occurs (Young et al., 1975). This results in deep-seated foundations causing local settlements under their edge. Large foundation bases resting on soil with high shear resistance lead to excessive deformation. Repeated loading from the waves in the lateral direction and excessive deformation result in softening of the soil particles in the vicinity of the foundation. This behavior is termed rocking.

After knowing the evolution of the structural form and the behavior of the structures, it is necessary to understand the encountered forces. For the analysis and design of offshore platforms, one has to determine the environmental forces acting on these structures, namely wind, current, wave, earthquake, ice forces, and so on. All the forces on a fixed or floating body can be categorized as small body or large body, based on the ratio of the characteristic dimension of the member or the structure with respect to the wavelength. For small bodies, the Morison equation is used, and for large bodies, diffraction–radiation wave theories are used to calculate the wave forces.

The various environmental forces acting on offshore structures have two main components; they vary both in space and time. They can be classified broadly into two, either on the basis of the physical phenomenon causing them, or on the uncertainties of the load. However, in the design, uncertainties are accounted for by using appropriate load factors. Major types of loads acting on offshore structures that could have dynamic effects are (i) functional loads arising from the operation of machinery and (ii) environmental loads due to wind, wave, current, ice, and earthquakes. In general, if the variation of these loads synchronizes with the natural period of the structure, it will result in dynamic amplification. Hence, load fluctuation with time can be classified into two parts, namely (i) macroscale variations that do not affect the structural response and (ii) microscale variations that affect the response significantly. Average wind velocity over a specified period of time (say 10 min), tidal current, significant wave height and the peak period of the wave spectrum, and peak ground acceleration of earthquake motion are examples of macroscale variations. In the case of environmental loads that show microscale variations, the variations are so rapid that they have a significant influence on the response of the structure. Microscale variations give rise to dynamic effects on structures.

1.6 WIND LOADS

Wind forces on offshore structures are caused by complex fluid-dynamics phenomena, which are generally difficult to calculate with high accuracy. The most widely used engineering approach to estimate wind forces on offshore structures is based on a few observations as listed below:

- When a stream of air flows with constant velocity (v), it will generate force on the flat plate of area (A).
- The plate will be placed orthogonal to the flow direction.
- This force will be proportional to (Av^2).
- The proportionality constant is independent of the area, which is verified by experimental studies.

Hence, the wind force on a plate, orthogonal to the wind flow direction can be determined by the net wind pressure as given below:

$$p_w = \frac{1}{2}\rho_a C_w v^2 \tag{1.1}$$

where ρ_a is the mass density of air (1.25 kg/m^3), and C_w is the wind pressure coefficient. It is important to note that the mass density of air increases due to the water spray (splash) up to a height of 20–20 m above MSL.

1.7 WAVE LOADS

Wind-generated sea surface waves can be represented by a combination of regular waves. Regular waves of different magnitude and wave lengths from different directions are combined to represent the sea surface elevation. Water particle kinematics of regular waves are expressed by the sea-surface elevation by various wave theories (Chandrasekaran and Bhattacharyya, 2012). Of all the theories, the Airy wave theory is commonly used because it assumes linearity between the kinematic quantities and the wave height, which makes it simple. The Airy theory assumes a sinusoidal waveform of wave height (H), which is small in comparison to the wave length (λ) and water depth (d) as given below:

$$
\begin{aligned}
\eta(x,t) &= \frac{H}{2}\cos(kx - \omega t) \\
k &= \frac{2\pi}{\lambda} \\
\dot{u}(x,t) &= \frac{\omega H}{2}\frac{\cosh(ky)}{\sinh(kd)}\cos(kx - \omega t) \\
\dot{v}(x,t) &= \frac{\omega H}{2}\frac{\sinh(ky)}{\sinh(kd)}\sin(kx - \omega t) \\
\ddot{u}(x,t) &= \frac{\omega^2 H}{2}\frac{\cosh(ky)}{\sinh(kd)}\sin(kx - \omega t) \\
\ddot{v}(x,t) &= -\frac{\omega^2 H}{2}\frac{\sinh(ky)}{\sinh(kd)}\cos(kx - \omega t)
\end{aligned}
\tag{1.2}
$$

The Airy theory is valid for mean sea level only. However, due to the variable submergence effect, the submerged length of the members will be continuously changing. This will attract additional forces due to their variable submergence at any given instant of time. To compute the water particle kinematics up to the actual level of submergence, stretching modifications suggested by various researchers are used.

1.8 MASS AND DAMPING

Mass is made up of the structural mass and the hydrodynamic added mass of the structure. For a slender structure, the mass of the displaced volume of the structure will be significant and should be considered in the analysis. Added mass depends on the submerged volume of the platform, which also varies with respect to the period of vibration. This is due to the variation in buoyancy, which in turn changes the tether tension variation that affects the natural frequency of motion. On the basis of equipment layout plan and the chosen structural form, one can compute the mass of the platform readily. It is also important to establish the fact that a desired proportion between center of buoyancy and center of mass is maintained to ensure stability under free floating conditions. This is important to enable smooth construction process in the case of floating structures.

For steel offshore structures, structural damping is usually considered to vary from 0.2 to 0.5 of that of the critical damping (Adams and Baltrop, 1991). For concrete structures, it can be of the order 0.5%–1.5%. Hydrodynamic damping originates from radiation damping and viscous damping effects. Radiation damping is determined using the potential theory. It exhibits a strong dependence on frequency and submergence effects. The literature shows that drag damping is lower for structures with large-diameter column members (~0.1%). The damping ratio for offshore structures (wet structures), including the effects of added mass, can be expressed as the ratio of dry ones.

1.9 ICE LOADS

Ice loads are dominant in offshore structures in the arctic regions. Prediction of ice loads is associated with a significant degree of uncertainty because there are various ice conditions that exist in the service life of an offshore platform. They are level ice, broken ice, ice ridges, and icebergs. Offshore structures show different types of failure under ice loads namely creep, cracking, buckling, spalling, and crushing. Ice loads exhibit random variations in both space and time. They are classified as (i) total or global loads and (ii) local loads or pressure. Global loads affect the overall motion and stability of the platform while local loads affect the members at connections. In the level ice condition, frequency of interaction between the structure and ice is important; the number of interactions per unit of time is important to quantify the ice loads on offshore platforms. Total ice force can result in a periodic loading and can cause dynamic amplification in flexible/slender structures. Current codes include equations for the extreme static ice loads, which depends on the geometric shape of the structure. Studies show that ice loads on a conical structure are less than those on a cylindrical structure. This is because a well-designed cone shape can change the ice-failure mode from crushing to bending. Estimating (predicting) ice forces on offshore platforms has a lot of uncertainties. Ice forces often control the design of the platform, particularly in operational conditions. The design ice loads use varying factors for level ice, first-year ridge ice, and multiyear ridge ice; the factored values are 2, 5, and 7, respectively.

1.10 SEISMIC LOADS

Offshore platforms that do not have a stiff connection with the seabed are indirectly influenced by earthquakes; those that are bottom-supported are affected by earthquakes directly. Compliant structures that are position-restrained by tethers will be subjected to dynamic tether tension variations under the presence of earthquake forces. This will give rise to dynamic tether tension variations, which in turn will affect the response of the platform under lateral loads. Earthquakes give rise to horizontal and vertical motions for a typical duration of 15–30 s. Earthquake acceleration exhibits random characteristics due to (i) the nature of the mechanism causing earthquakes; (ii) wave propagation; (iii) reflection, and (iv) deflection. The effects of earthquake forces give rise to horizontal and vertical motions with durations of 15–30 s. Earthquake loads exhibit random characteristics due to the nature of the mechanism causing earthquake, wave propagation, reflection, and deflection. Earthquakes can result in inertia forces due to the acceleration and damping forces due to the motion of the water particles.

1.11 CURRENT FORCES

The presence of currents in water produces the following distinct effects:

- Current velocity should be added vectorially to the horizontal water particle velocity before computing the drag force because drag force depends on the square of the water particle velocity. Current decreases slowly with increase in depth but even a small magnitude of current velocity can cause significant drag force.
- Steepening of the wave profile by changing the wave celerity. This effect is insignificant and generally neglected.
- Currents cause the structure itself to generate waves, which in turn create diffraction forces. However, these values are negligible for realistic values of currents acting on normal-sized members.
- The presence of currents is alternatively accounted for by increasing the wave height by 10%–15% and neglecting the presence of current per se.

1.12 DEAD LOAD

Dead load, which includes pilings, superstructure, jacket, stiffeners, piping, conductors, corrosion anodes, deck, railing, grout, and other appurtenances, is the weight of the overall platform in air. Dead load excludes the following:

1. Weight of the drilling equipment placed on the platform including the derrick, drawworks, mud pumps, mud tanks, and so on.
2. Weight of production or treatment equipment located on the platform including separators, compressors, piping manifolds, and storage tanks.
3. Weight of drilling supplies that cause variable loads during drilling such as drilling mud, water, fuel, casing, and so on.
4. Weight of treatment supplies employed during production such as fluid in the separator and storage in the tanks.
5. Drilling load, which is approximate combination of derrick load, pipe storage, rotary table load, and so on.

1.13 LIVE LOADS

Live loads act in addition to the equipment loads. They include loads caused by impacts of vessels and boats on the platform. A dynamic amplification factor is applied to such loads to compute the enhanced live loads. Live loads are generally designated as a factor times the applied static load. These factors are assigned by the designer depending on the type of platform.

1.14 DESIGN REQUIREMENTS

The design methodology for offshore platforms differs with different types of offshore structures. For example, vertical deformation will be less in the case of bottom-supported structures such as jacket platform, GBS, and so on. Such platforms are highly rigid and tend to attract more forces. Hence the design criterion should be to limit the stresses in the members. Displacement of the members under the applied loads will be insignificant. On the other hand, compliant structures are more flexible because they all displace more under wave action. They also create more disturbances in the waves. Hence the design criterion will be control of displacement instead of limiting the stresses in the members. Orientation of the platform is another important aspect in the design. It is preferred that members are oriented to have less projected area in the encountered wave direction. This induces a lesser response in the members. The predominant wave direction for the chosen site is made available to the designer, and based on this information the platform orientation is decided (Chandrasekaran and Bhattacharyya, 2012). The following is a list of data required for the design of offshore structures:

- Land topographical survey of sufficient area covering the chosen site for platform installation
- Hydrographical survey of the proposed location. Hydrographic charts are used for this purpose
- Information regarding silting at the site
- Wind rose diagram showing information on wind velocities, duration, predominant direction year-round
- Cyclonic tracking data showing details of the past cyclonic storms, such that wind velocities, direction, peak velocity period, and so on, are indicated
- Oceanographic data including general tide data, tide tables, wave data, local current, seabed characteristics, temperature, rainfall, and humidity
- Seismicity level and values of acceleration
- Structural data of existing similar structures, preferably in the near vicinity
- Soil investigation report

The analysis of an offshore structure is an extensive task, embracing consideration of the different stages—execution, installation, and in-service stages—during its life. Many disciplines, for example, structural geotechnical engineering, naval architecture, and metallurgy are involved. The analytical models used in offshore engineering are in some respects similar to those adopted for other types of steel structures. Only the salient features of offshore models are presented here. The same model is used throughout the analysis process with only minor adjustments being made to suit the specific conditions, particularly, for example, at supports relating to each analysis. Stick models (beam elements assembled in frames) are used extensively for tubular structures (jackets, bridges, flare booms) and lattice trusses (modules, decks). Each member is (normally) rigidly fixed at its ends to other elements in the model. If more accuracy is required, particularly for the assessment of natural vibration modes, local flexibility of the connections may be represented by a joint stiffness matrix. In addition to its geometrical and material properties, each member is characterized by hydrodynamic coefficients, for example relating to drag, inertia, and marine growth to allow wave forces to be automatically generated. The integrated decks and hulls of floating platforms involving large bulkheads are described by plate elements. The characteristics assumed for the plate elements depend on the principal state of stress to which they are subjected. Membrane stresses are taken when the element is subjected merely to axial load and shear. Plate stresses are adopted when bending and lateral pressure are to be taken into account. After developing a preliminary model for analysis, member stresses are checked for the preliminary sizing under different environmental loads.

The verification of an element consists of comparing its characteristic resistance(s) to a design force or stress. It includes namely (i) a strength check in which the characteristic resistance is related to the yield strength of the element and (ii) a stability check for elements in compression in which the characteristic resistance relates to the buckling limit of the element. An element (member or plate) is checked at typical sections (at least both ends and midspan) against resistance and buckling. This verification also includes the effect of water pressure for deepwater structures. Tubular joints are checked against punching under various load patterns. These checks may indicate the need for local reinforcement of the chord using overthickness or internal ring-stiffeners. Elements should also be verified against fatigue, corrosion, temperature, or durability wherever relevant.

1.15 FABRICATION AND INSTALLATION LOADS

These loads are temporary and arise during fabrication and installation of the platform or its components. During fabrication, various structural components generate lifting forces, while in the installation phase forces are generated during platform loadout, transportation to the site, launching, and upending, as well as during lifts related to installation. According to the DNV rules, the return period for computing design environmental conditions for installation and fabrication loads is three times that of the duration of the corresponding phase. API-RP2A, on the other hand, leaves this design return period up to the owner, while the BS6235 rules recommend a minimum recurrence interval of 10 years for the design environmental loads associated with transportation of the structure to the offshore site.

1.16 LIFTING FORCES

Lifting forces are functions of the weight of the structural component being lifted, the number and location of lifting eyes used for the lift, the angle between each sling and the vertical axis, and the conditions under which the lift is performed. All members and connections of a lifted component must be designed for the forces resulting from static equilibrium of the lifted weight and the sling tensions. Moreover, API-RP2A recommends that in order to compensate for any side movements, lifting eyes and the connections to the supporting structural members should be designed for the combined action of the static sling load and a horizontal force equal to 5% this load applied perpendicular to the padeye at the center of the pinhole. All these design forces are applied as static loads if the lifts are performed in the fabrication yard. If, however, the lifting derrick or the structure to be lifted is on a floating vessel, then dynamic load factors should be applied to the static lifting forces. A factor of 2 is applied for members and connections and 1.35 for all other secondary members. For loadout at sheltered locations, the corresponding minimum load factors for the two groups of structural components are 1.5 and 1.15, respectively.

1.17 LOADOUT FORCES

These are forces generated when the jacket is loaded from the fabrication yard onto the barge. If the loadout is carried out by direct lift, then, unless the lifting arrangement is different from that to be used for installation, lifting forces need not be computed. This is due to the fact that lifting in the open sea creates a more severe loading condition, which requires higher dynamic load factors. If loadout is done by skidding the structure onto the barge, a number of static loading conditions must be considered, with the jacket supported on its side. Such loading conditions arise from the different positions of the jacket during the loadout phases. Because the movement of the jacket is slow, all loading conditions can be taken as static. Typical values of friction coefficients for the calculation of skidding forces are namely (i) steel on steel without lubrication (0.25); (ii) steel on steel with lubrication (0.15); (iii) steel on teflon (0.10); and (iv) teflon on teflon (0.08).

1.18 TRANSPORTATION FORCES

These forces are generated when platform components (jacket, deck) are transported offshore on barges or self-floating. They depend on the weight, geometry, and support conditions of the structure (by barge or by buoyancy) and also on the environmental conditions (waves, winds, and currents) that are encountered during transportation, that is, the types of motion that a floating structure may experience. To minimize the associated risks and secure safe transport from the fabrication yard to the platform site, it is important to plan the operation carefully by considering the following (API-RP-2A):

- Previous experience along the tow route
- Exposure time and reliability of predicted "weather windows"
- Accessibility of safe havens
- Seasonal weather system
- Appropriate return period for determining design wind, wave, and current conditions, taking into account characteristics of the tow such as size, structure, sensitivity, and cost

1.19 LAUNCHING AND UPENDING FORCES

These forces are generated during the launch of a jacket from the barge into the sea and during the subsequent upending into its proper vertical position to rest on the seabed. The five stages of jacket motion in a launch-upending operation are (i) the jacket slides along the skid beams; (ii) rotates on the rocker arms; (iii) rotates and slides simultaneously; (iv) detaches completely and comes to its floating equilibrium position; and (v) is upended by a combination of controlled flooding and simultaneous lifting by a derrick barge. Both the static and dynamic loads for each stage of the above under the action of wind, waves, and current need to be included in the analysis. To start the launch, the barge must be ballasted to an appropriate draft and trim angle and subsequently the jacket must be pulled toward the stern by a winch. Sliding of the jacket starts as soon as the downward force (gravity component and winch pull) exceed the friction force. Because the jacket slides, its weight is supported on the two legs that are part of the launch trusses. The support length keeps decreasing and reaches a minimum, equal to the length of the rocker beams, when rotation starts. It is generally at this instant that the most severe launching forces develop as reactions to the weight of the jacket. During the last two stages, variable hydrostatic forces arise, which have to be considered for all members affected. Buoyancy calculations are required for every stage of the operation to ensure fully controlled, stable motion. Computer programs are available to perform the stress analyses required for launching and upending and also to portray the whole operation graphically.

1.20 ACCIDENTAL LOADS

According to the DNV rules, accidental loads are ill-defined with respect to intensity and frequency, which may occur as a result of accident or exceptional circumstances. Examples of accidental loads are loads due to collision with vessels, fire or explosion, dropped objects, and unintended flooding of buoyancy tanks. Special measures are normally taken to reduce the risk from accidental loads. For example, protection of wellheads or other critical equipment from a dropped object can be provided by specially designed, impact-resistant covers. An accidental load can be disregarded if its annual probability of occurrence is less than 10^{-4}. This number is meant as an order of magnitude estimate and is extremely difficult to compute.

2 Ultimate Load Design

2.1 INTRODUCTION

The fundamental objective of a structural design procedure is to address both the load resistance and intended functionality along with the margin of safety (Freudenthal and Gaither, 1969; Abrahamsen, 1976). It is necessary to ensure that the chosen structural geometry must also fit the serviceability criteria during normal use. The limit state of design for marine structures is explained in terms of the following limit states:

a. The ultimate limit state refers to the maximum load-carrying capacity with excessive yielding, which transforms the structure into a mechanism. Buckling may occur due to elastic or elastic–plastic instability, which will lead to loss of equilibrium of the complete structure.
b. The fatigue limit state indicates damage caused by loads that are cyclic in nature. This includes stress concentration, which occurs due to the reversal nature of stresses. Cyclic loading arise generally from waves, wind, currents, live loads, machinery, etc.
c. The progressive limit state refers to the limit state of progressive collapse, which is caused by progressive damage.
d. The serviceability limit state corresponds to the normal use and durability of the structure.

Steel structures behave in a specific manner such that the design concepts differ from conventional methods. This is because steel has a unique physical property—ductility. Steel possesses higher reserve strength beyond the yield point. The ultimate load method utilizes the full reserve strength. A lot of uncertainties that arise from material properties, construction methods, and design procedures make the design probabilistic rather than deterministic. It is therefore important to address these uncertainties for improved accuracy in the design (Love, 1994). The uncertainties that are responsible for affecting the design principle are as follows:

a. Variability in material quality with respect to strength and physical properties (E, σ)
b. Variability in geometric properties of members, for example, diameter and thickness
c. Uncertainties in the specified load with respect to environmental conditions, for example, wind
d. Idealization of the real structure to a mathematical model
e. Errors that arise from the numerical methods used in analysis and design

f. Quality of workmanship
g. Material degradation due to corrosion

Uncertainties due to the above factors are evident, but they become significant in the design only when one needs to use the material strength to the ultimate or maximum level.

Ultimate limit design accounts for these uncertainties up to a certain point by means of safety factors. A safety factor in the design process is based on the probabilistic method, with an understanding of variabilities/uncertainties and their permissible limits. Ultimate limit design handles this issue in a semiprobabilistic manner for the strength limit, beyond which variables are not acceptable. Structural members should be chosen to meet the following requirements:

$$\frac{\text{Characteristic strength}}{\text{Factor of uncertainty}} \geq (\text{Characteristic load} * \text{Factor of uncertainty})$$

2.2 IDEALIZED STRESS–STRAIN CURVE

A stress–strain curve for mild steel is shown in Figure 2.1. The material remains elastic up to the elastic limit and obeys Hooke's law.

Hooke's law states that within the elastic limit, stress is proportional to strain. Since for most materials it is difficult to express the entire stress–strain curve with a simple mathematical expression, the behavior of the materials is generally represented by an idealized stress–strain curve (Figure 2.2).

From Figure 2.2, one can infer that until the point "A" the material obeys Hooke's law. Point "A" is termed the *proportionality limit,* beyond which variations are nonlinear. Point "B" represents the *elastic limit,* and is the maximum stress value up to which the material remains elastic; it will regain its original shape and size on removal of the externally applied load. Point "C" represents the *upper yield point,* and "D" represents the *lower yield point,* where the strain increases without any increase in the stress. It is rather difficult to obtain the stress–strain relationship in this region as it coincides with the yield point; proportionality and elastic limits are concluded in this region. Further beyond this point, both stress and strain

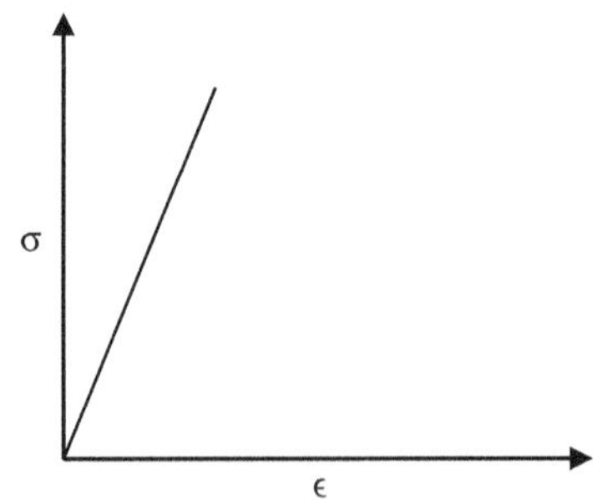

FIGURE 2.1 Linearly elastic material.

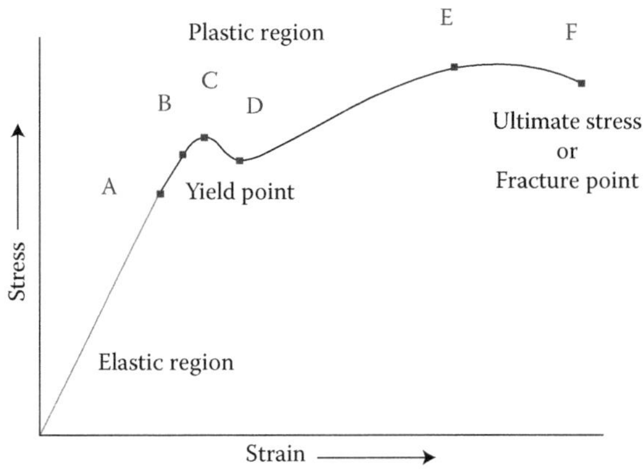

FIGURE 2.2 Stress–strain curve for steel.

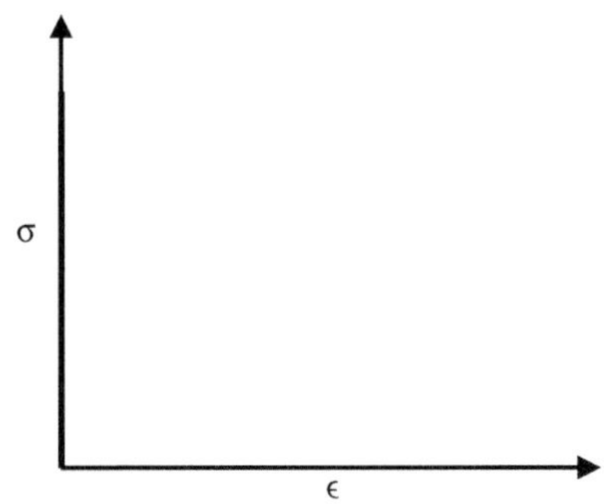

FIGURE 2.3 Stress–strain curve for rigid materials.

increase considerably in the *strain hardening region*. When the material reaches the maximum stress value, there is a dip in the stress until it reaches the breaking point. Plastic theory actually involves taking even the strain-hardening region into design considerations. Figures 2.3 through 2.5 show the stress–strain curve for different materials.

It is reiterated that the principal objective of structural design is to ensure safe load-carrying capacity in addition to the serviceability requirements. In plastic design, it is extended by considering the ultimate capacity of the material/member.

2.3 PLASTIC ANALYSIS

Consider a simply supported beam subjected to an external load. When the load is increased, the moment is also increased proportionally, until the extreme fiber of the member is stressed to the yield point. Beyond the yield point, any further increase in load will result in a significant permanent deformation of the material,

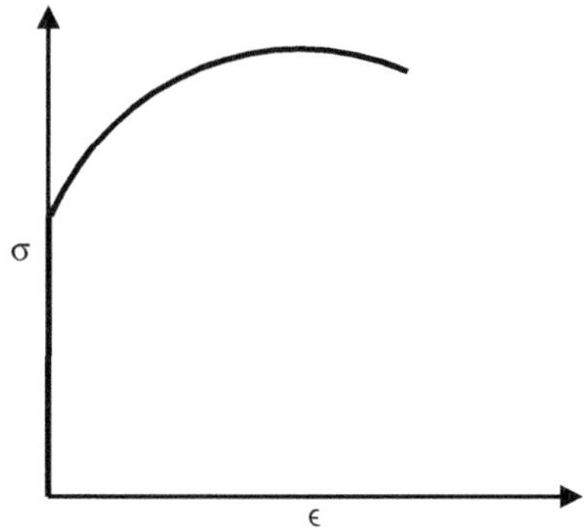

FIGURE 2.4 Stress–strain curve for rigid plastic materials (strain hardening).

which in turn will increase the deflection significantly. Therefore, the load causing the first yield point is said to be the critical load, as this affects the serviceability conditions (by leading to unacceptable large deflections). However, the phenomenon varies according to the boundary conditions; when we consider a beam with fixed boundary conditions, it behaves differently—the limit of proportionality falls, as shown in Figure 2.6.

Because of the ductility of steel, the moment capacity that is developed at the first zone of yielding remains the same, irrespective of the boundary conditions. Further increase in load increases the zone of yielding and makes the member inelastic. Deflection increases without any significant increase in the load-carrying capacity. Hence, failure of the member does not correspond to the attainment of the elastic limit, but to the reserve strength beyond the yield point. This is much higher than the load at the yield point. This emphasizes a basic fact that redundant structures can be effectively designed using plastic analysis due this higher reserve strength. The plastic method of analysis is generally applied to statically indeterminate structures, where the joints transfer bending moments from the highly stressed sections to other sections.

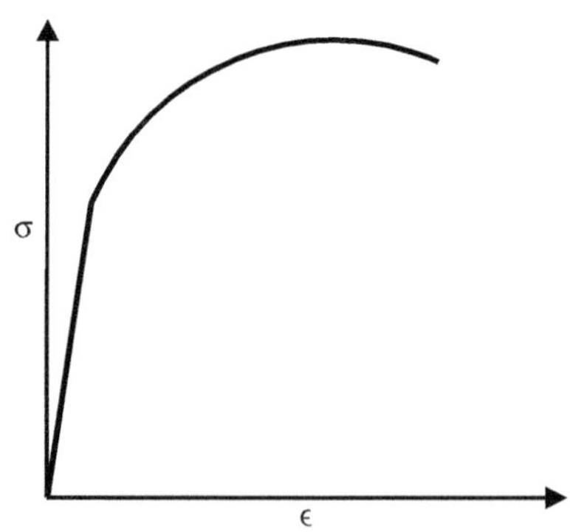

FIGURE 2.5 Stress–strain curve for elastic–plastic material.

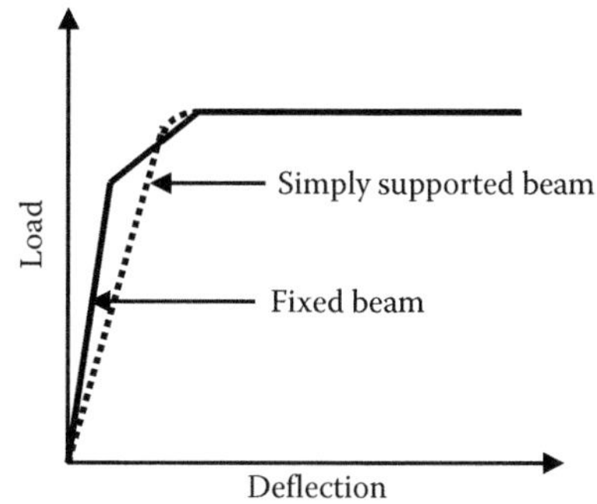

FIGURE 2.6 Load–deflection curve for simply supported and fixed beams.

2.4 ULTIMATE LOAD-CARRYING CAPACITY

2.4.1 Tension Member

Let us consider a member with cross-sectional area A subjected to axial tension P. Figure 2.7 shows the load–deflection curve for this member. Deflection at the elastic limit (δ_y) at the corresponding load (P_u) is given by

$$\delta_y = \left(\frac{P_u l}{AE}\right) \tag{2.1}$$

where E is the modulus of elasticity of the material, l is the unsupported length, and A is the cross-sectional area of the member.

The plastic flow of stress distribution starts within the member when the load reaches the value $P_u(=f_y A_s)$, and this value remains constant and is called the ultimate load for a tension member. The ultimate load concept is of little use in the design of tension members, as it is obvious from Figure 2.7 that the section is damaged as a

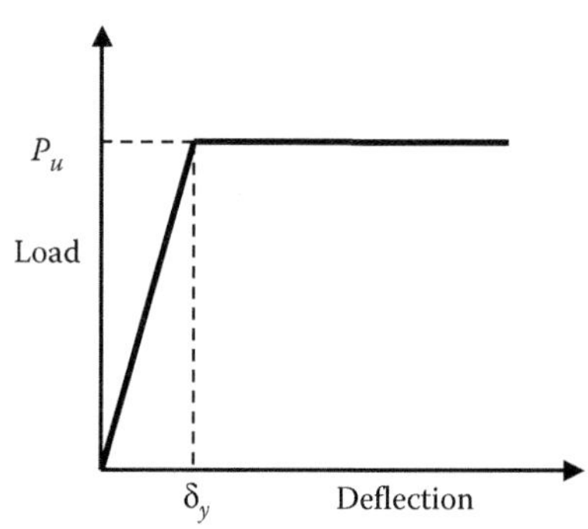

FIGURE 2.7 Load–deflection curve for tension members.

result of a large deflection. The failure stress is almost twice that of the elastic limit in tension with reference to the original cross-sectional area. This infers that there is higher reserve capacity in a tension member.

2.4.2 Flexural Member

The concept used in the plastic analysis of a flexural member is explained by considering the beam with the centroid "*O*," having an axis of symmetry as shown in Figure 2.8.

Material is assumed to obey the ideal elastic–plastic stress–strain relationship. The beam is assumed to deflect due to a pure couple and the effect of axial and shear forces is neglected. Deformation and strain are considered to be minimum, so the longitudinal stresses are neglected. Further, it is also assumed that the plane section remains plane even after bending. The cross section is assumed to be symmetric with respect to the axis in the plane of bending. When the stress value in the extreme fibers in both tension and compression reaches yield, the corresponding strain reaches the yield value. Any further increase in the moment increases the strain in further fibers in the member. Hence, the neutral axis no longer passes through the centroid and is shifted to maintain force equilibrium, which will be renamed as the equal area axis (EAA). This implies that, about this axis, total compressive and tensile force remain equal, resulting in moment equilibrium. Yielding of the cross section spreads inwards from the extreme stressed fiber. On further increase in the moment, the entire section will tend to yield. This will ensure that the stress in all fibers, both in the tension and the compression zones, will be equal to the yield value. Any further increases in moment beyond this state will result in rotation, as no additional resisting moment will be possible. The moment-carrying capacity of the section at this stage is termed the plastic moment of resistance (M_p), which is given by the product of the force (either tensile or compressive; the product of the force does not matter because the section is already in force equilibrium about the EAA) and the lever arm distance between the centroid of the forces.

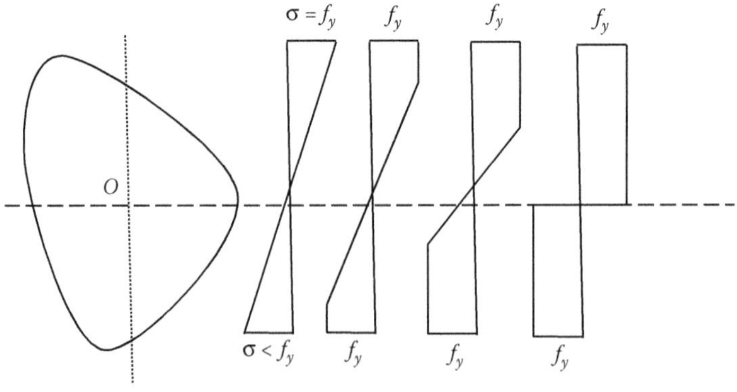

FIGURE 2.8 Development of full plasticity of a beam section.

2.5 PLASTIC HINGE

A plastic hinge is a zone of yielding due to flexure in the structural member. Although hinges do not actually form, it can be seen as the slope occurs over the smaller length of the member at the sections of the maximum moments. Strain hardening occurs in these regions, as there is a significant increase in the deflection with a slight increase in the load. Members remain elastic until the moment reaches the maximum moment of resistance of the fully yielded cross section, which is the plastic moment capacity of the section. Further additional moment will cause rotation with the increase in stress. Hence, a plastic hinge can be redefined as the yielded zone in a member in which infinite rotation occurs at a constant restraining moment (M_p). The analysis is carried out by assuming the discrete points of zero length at critical sections, where the plastic hinge formed first reaches the maximum deformation. Ideal sections are primarily: (i) sections where loads are concentrated, (ii) sections where the cross section in a member changes, (iii) sections where zero shear in the span under distributed load occurs, and (iv) at fixed supports. Classical differences between the plastic hinge and the structural hinge (hinged joints) are as follows:

i. Plastic hinges have a moment-carrying capacity of M_p, whereas structural hinges have no moment capacity.
ii. Plastic hinge locations are marked as filled circles, whereas structural hinges are represented as unfilled circles.
iii. Plastic hinges have limited rotation capacity, whereas structural hinges have unlimited rotation capacity.

In simple terms, plastic hinges are equivalent to rusted structural hinges, which require moment to rotate them (M_p).

2.6 HINGE LENGTH

Let us consider a simply supported beam of rectangular cross section, which is subjected to a gradually increased concentrated central load, as shown in Figure 2.9. In the central section—the highly stressed cross section—a plastic hinge will format the center, as shown in the figure.

$$M_p = \left(\frac{Pl}{4}\right) \tag{2.2}$$

$$M_y = f_y Z = f_y\left(\frac{bh^2}{6}\right) = \frac{2}{3} f_y\left(\frac{bh^2}{4}\right) = \frac{2}{3} M_p \tag{2.3}$$

From the bending moment diagram, the following can be deduced:

$$\frac{M_p}{L/2} = \frac{M_y}{L/2 - x/2}$$

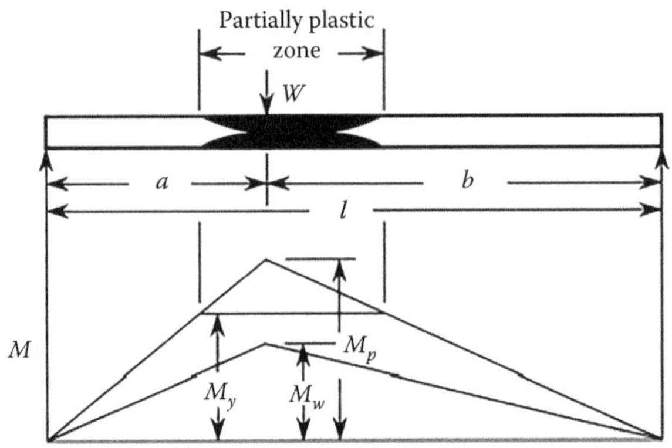

FIGURE 2.9 Development of plastic hinge and hinge length.

$$\frac{(L-x)}{2}M_p = \frac{L}{2}M_y$$

$$(L-x)M_p = LM_y$$

$$(L-x)M_p = L\left(\frac{2}{3}\right)(M_p) \tag{2.4}$$

$$(L-x) = \frac{2}{3}L$$

$$x = \frac{L}{3} \tag{2.5}$$

Therefore, the hinge length of the plasticity zone is equal to one-third that of the span.

2.7 PLASTIC MOMENT OF A SECTION

To derive the plastic moment of resistance whose cross section is fully plasticized, the following assumptions are made:

a. The material obeys Hooke's law, until stress reaches the first yield value.
b. The upper and lower yield points in tensile and compressive stress are the same.
c. The material is homogeneous and isotropic in both the elastic and plastic states.
d. A plane transverse section (a section that is perpendicular to the longitudinal axis of the member) will remain plane and normal, even after bending.

e. There is no resultant force acting on the beam.
f. The cross section is symmetric about the centroidal axis and is parallel to the plane of bending.
g. Every layer of the cross section is free to expand and contract and remains independent with respect to the adjacent layer.

For a fully plastic section, it is mandatory that the section maintains a force equilibrium about the EAA. For a symmetric section, as assumed in the present case, EAA will coincide with that of the centroidal axis, and hence the following holds good:

Total compressive force (C) = Total tensile force (T)

$$\sigma_y A_1 = \sigma_y A_2$$

$$A_1 = A_2 = \frac{A}{2}$$

$$M = C\bar{y}_1 + T\bar{y}_2$$

$$M = \sigma_y A_1 \bar{y}_1 + \sigma_y A_2 \bar{y}_2$$

$$= \sigma_y \frac{A}{2}(\bar{y}_1 + \bar{y}_2)$$

$$M_p = \sigma_y Z_p \tag{2.6}$$

where Z_p is the plastic section modulus and σ_y is the yield stress. For the stress distribution shown in Figure 2.10, the plastic moment of resistance can be rewritten as

$$M_P = \frac{\bar{A}}{2}(\bar{y}_1 + \bar{y}_2) \tag{2.7}$$

$$M_p = M_{ult} = \sigma_y Z_p \tag{2.8}$$

$$M_e = \sigma_y Z_e \tag{2.9}$$

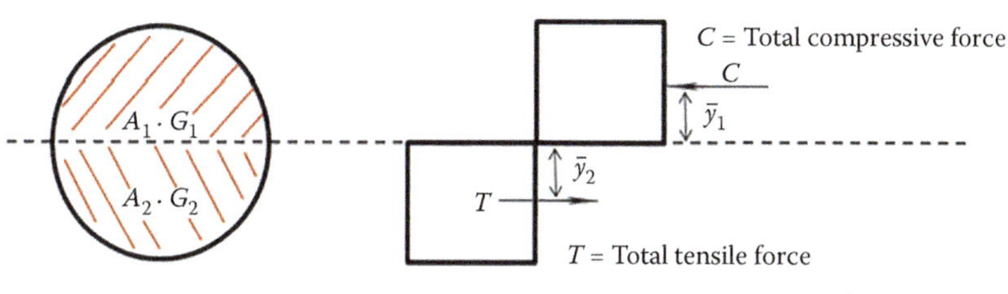

FIGURE 2.10 Plastic moment of a section.

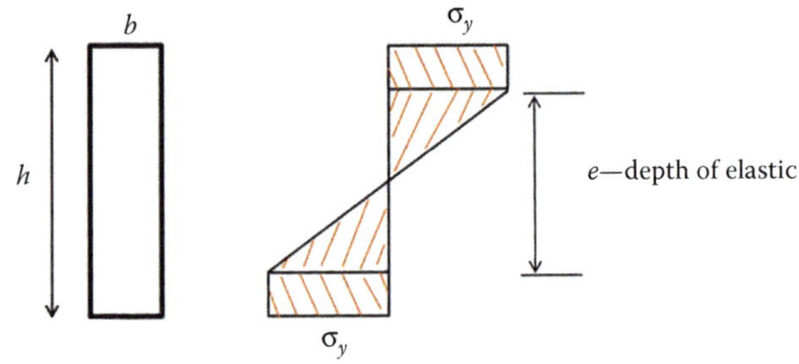

FIGURE 2.11 Moment capacity of an elastic–plastic section.

Now, consider a rectangular section, whose stress distribution is elastic–plastic, as shown in Figure 2.11. Let M_1 and M_2 be moment capacities of elastic and plastic sections, respectively. Then, the total moment capacity of the elastic–plastic section is given by

$$M = M_1 = M_1 + M_2$$

To find M_1, the following relationship is valid:

$$M_1 = \sigma_y \left\{ \left(\frac{1}{2} b \frac{e}{2} \right) \left(\frac{2}{3} \frac{e}{2} \right) * 2 \right\}$$

$$M_1 = \sigma_y \frac{be^2}{6}$$

To find M_2, the following relationship is valid:

$$\left(\frac{bh}{2} \right) \left(\frac{h}{4} \right)^2 = \frac{bh^2}{4} \tag{2.10}$$

For a fully plastic section, the following relationship holds good:

$$Z = \left(\frac{be}{2} \right) \left(\frac{e}{4} \right) (2) = \frac{be^2}{4}$$

$$M_2 = \sigma_y \left\{ \frac{bh^2}{4} \frac{be^2}{4} \right\} = M_1 + M_2$$

$$M = \sigma_y \left\{ \frac{be^2}{6} \frac{bh^2}{4} \frac{be^2}{4} \right\} = \sigma_y \left\{ \frac{bh^2}{4} - \frac{be^2}{4} \right\} = M = \sigma_y \frac{bh^2}{4} \left(1 - \frac{e^2}{3h^2} \right)$$

$$Z_p = \frac{A}{2}(\bar{y}_1 + \bar{y}_2) = \frac{bh}{2}\left(\frac{h}{4} + \frac{h}{4}\right) = \frac{bh^2}{4}$$

$$M = \sigma_y \frac{bh^2}{4}\left(1 - \frac{e^2}{3h^2}\right)$$

$$M = \sigma_y Z_p \left(1 - \frac{e^2}{3h^2}\right)$$

$$M = M_p \left(1 - \frac{e^2}{3h^2}\right) \tag{2.11}$$

For the above relationship, one can find the depth of the elastic core of the elastic–plastic cross section for known values of moment acting on the section. The ratio of the plastic and elastic moments is known as the shape factor and is given by

$$\frac{M_p}{M_e} = S = \frac{\sigma_y z_p}{\sigma_y z_e} = \frac{Z_p}{Z_e} = \text{shape factor} \tag{2.12}$$

2.8 SHAPE FACTOR FOR DIFFERENT CROSS SECTIONS

The shape factor is defined as the ratio of plastic moment to the yield moment of the section. It is a function of the cross section form or shape, hence the name *shape factor.*

2.8.1 Rectangular Cross Section

A typical rectangular section whose stress distribution is shown in Figure 2.12 has a depth of the elastic core as e.

$$s = \frac{Z_p}{Z_e}$$

$$Z_p = \frac{A}{2}(\bar{y}_1 + \bar{y}_2) = \frac{bh}{2}\left\{\frac{h}{4} + \frac{h}{4}\right\} = \frac{bh^2}{4}$$

$$Z_e = \frac{I}{y_m} = \frac{bh^3}{12} * \frac{1}{\frac{h}{2}} = \frac{bh^2}{6}$$

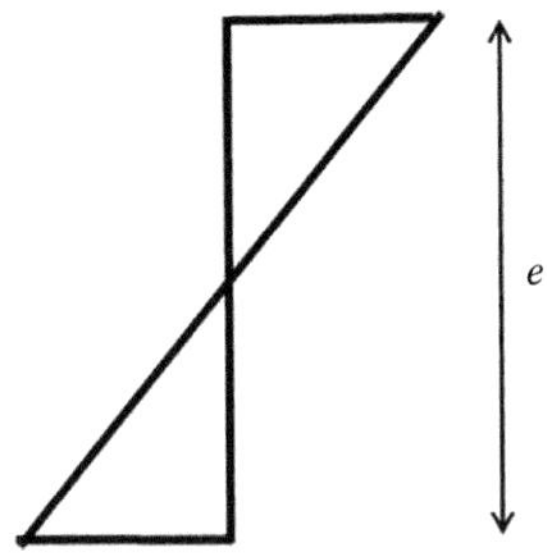

FIGURE 2.12 Rectangular section with elastic core.

The shape factor is given by

$$s = \frac{Z_p}{Z_e} = \frac{bh^2}{4} * \frac{6}{bh^2} = 1.5$$

$$M_p = 1.5M_e = 1.5(\sigma_y Z_e)$$

2.8.2 Circular Cross Section

A typical circular cross section is shown in Figure 2.13. For the assumed section, the centroid of the compressive and tensile forces with respect to EAA is given by

$$\bar{y}_1 = \frac{4r}{3\pi}; \quad \bar{y}_2 = \frac{4r}{3\pi}$$

The shape factor is given by

$$s = \frac{Z_p}{Z_e}$$

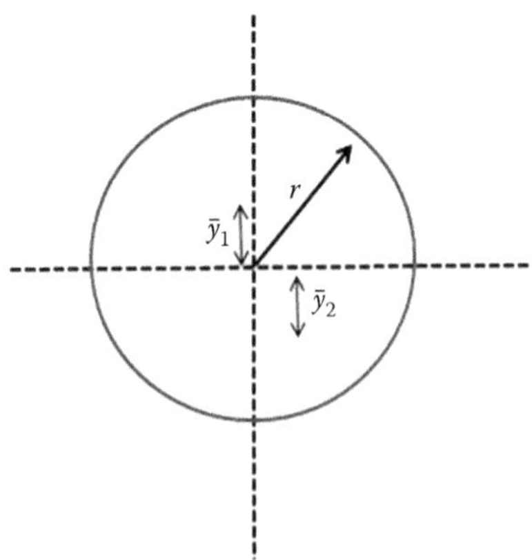

FIGURE 2.13 Circular section.

$$Z_p = \frac{A}{2}(\bar{y}_1 + \bar{y}_2) = \frac{\pi r^2}{2}\left\{\frac{4r}{3\pi}\frac{4r}{3\pi}\right\}$$

$$Z_p = \frac{4r^3}{3}$$

$$z_e = \frac{I}{y_m} = \frac{I}{r}$$

$$= \frac{\pi d^4}{64}\frac{1}{2}$$

$$= \frac{\pi}{64}(d)^4\frac{1}{2}$$

$$= \frac{\pi}{64}2^4\frac{r^4}{8} = \frac{\pi r^3}{4}$$

$$s = \frac{Z_p}{Z_e} = \frac{4r^3}{3} * \frac{4}{\pi r^3} = \frac{\pi r^3}{4} = \frac{16}{3\pi} = 1.69$$

2.8.3 Tubular Cross Section

A typical tubular cross section is shown in Figure 2.14.

The moment of inertia about the major axis is given by

$$I_{xx} = \frac{\pi}{64}(D_1^4 - D_2^4) = \frac{\pi}{4}[(2r_1)^4 - (2r_2)^4]$$

$$= \frac{\pi}{4}(r_1^4 - r_2^4)$$

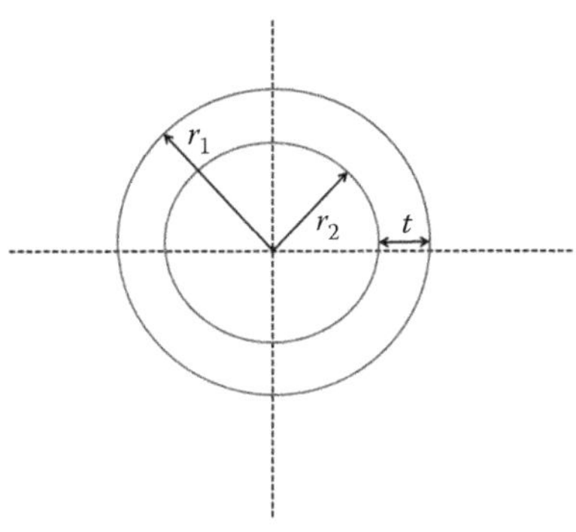

FIGURE 2.14 Tubular cross section.

The elastic section modulus is given by

$$Z_e = Z_r = \frac{I_{xx}}{r_1} = \frac{\pi}{4}\left[\frac{r_1^4 - r_2^4}{r_1}\right]$$

$$\bar{y}_1 = \frac{\varepsilon A \bar{y}}{\varepsilon A}$$

$$\bar{y}_1 = \frac{(\pi r_1^2/2)(4r_1^2/2) - (\pi r_2^2/2)(4r_2/3\pi)}{(\pi/2)(r_1^2 - r_2^2)}$$

$$= \frac{(2/3)(r_1^3 - r_2^3)}{(\pi/2)(r_1^2 - r_2^2)}$$

$$= \frac{4}{3\pi}\left[\frac{r_1^3 - r_2^3}{r_1^2 - r_2^2}\right]$$

$$Z_p = \frac{A}{2}(\bar{y}_1 + \bar{y}_2)$$

$$= \frac{\pi}{2}(r_1^2 - r_2^2)2\left\{\frac{4}{3\pi}\left[\frac{r_1^3 - r_2^3}{r_1^2 - r_2^2}\right]\right\}$$

$$= \frac{4}{3}(r_1^2 + r_2^2)$$

The shape factor is given by the following expression:

$$\frac{r_2}{r_2} = K = \frac{16r_1\ r_1^3(1-K^3)}{3\pi\ r_1^4(1-K^4)} = \frac{16}{3\pi}\left(\frac{1-K^3}{1-K^4}\right) \tag{2.13}$$

for $r_2 = 0$, the shape factor, as obtained from the above expression, is 1.698 for a solid circular cross section. Please note that this is the same as that obtained from Section 2.8.2.

2.9 FRAMED STRUCTURES

For framed structures, as shown in Figure 2.15, it is convenient to use the following relationship:

$$\frac{M_p}{M_y} = s$$

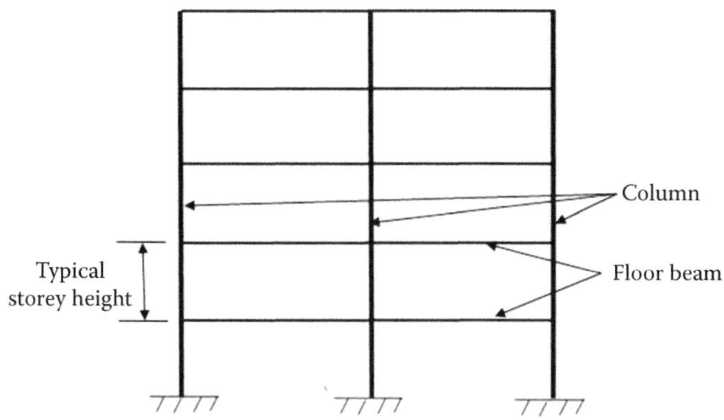

FIGURE 2.15 Framed structure.

$$M_y = \sigma_y$$

$$M_p = \sigma_y Z_p$$

$$Z_p = \frac{A}{2}(\bar{y}_1 + \bar{y}_2)$$

2.9.1 Moment Curvature Relationship

For simple bending problems, the following equation holds good:

$$\frac{M}{I} = \frac{\sigma}{y} = \frac{E}{R}$$

$$\frac{1}{R} = \frac{M}{EI} = \Phi$$

Further,

$$\frac{E}{R} = \frac{\sigma}{y}$$

$$\frac{E}{R} = \frac{2\sigma_y}{h} = \frac{E}{K}$$

$$\frac{1}{R} = \frac{2\sigma_y}{Eh} \tag{2.14}$$

Let e denote the depth of the elastic core in a known cross section. The section is likely to become completely plasticized when the moment is increased beyond a limit, which results in the bending stress exceeding the yield stress value. Hence, the following relationship holds good:

$$\frac{1}{R} = \frac{\sigma_y}{E}$$

$$\frac{1}{R} = \Phi_y = \left(\frac{M}{EI}\right)_{yield}$$

$$\Phi_y = \frac{M}{EI}$$

$$\Phi_p = \frac{M_p}{EI}$$

$$\frac{M_p}{M_y} = \frac{\Phi_p}{\Phi_y} = S$$

where s is the shape factor. Then the above expression is rewritten as

$$\frac{\sigma_y Z_p}{\sigma_y Z_y} = \frac{\Phi_p}{\Phi_y} = S$$

The following relationships also hold good:

$$\text{Also, } M = M_p\left(1 - \frac{e^2}{3h^2}\right)$$

$$\frac{M}{M_p} = 1 - \frac{e^2}{3h^2}$$

$$\frac{1}{R} = \frac{2\sigma_y}{E}$$

Substituting, we get

$$e = \frac{R2\sigma_y}{E}$$

$$\frac{M}{M_p} = \left\{1 - \left[\left(\frac{2p\sigma_y}{E}\right)^2 \frac{1}{3h^2}\right]\right\}$$

$$= \left\{1 - \frac{1}{3}\left(\frac{2\sigma_y}{E}\right)^2 \frac{1}{(h/R)^2}\right\} \tag{2.15}$$

The following relationship is also valid:

$$\frac{1}{R} = \frac{2\sigma_y}{Eh}$$

$$\frac{2\sigma_y}{E} = \left(\frac{h}{R}\right)_{yield} \tag{2.16}$$

Rewriting, we get

$$\frac{M}{M_p} = \left\{1 - \frac{1}{3}\left(\frac{h}{R}\right)_y^2 \frac{1}{(h^2/R)}\right\}$$

$$\frac{M}{M_p} = \left\{1 - \frac{1}{3}\left[\frac{h/R_y}{h/K}\right]\right\} \tag{2.17}$$

In terms of design perspective, the following relationship will be useful:

$$\frac{M}{M_p} = \left[1 - \frac{e^2}{3h^2}\right] \tag{2.18}$$

We know that the following relationship is valid:

$$M_y = 2p\sigma_y$$

$$= (Z * S)\sigma_y$$

$$= (\text{section modulus * shape factor})\sigma_y$$

$$\text{Load factor} = \frac{w_c}{w_w}$$

where w_c is the collapse load and w_w is the working load.

In plastic design, yield stress is assumed to be constant, and the allowable stresses are considered only as a fraction of the yield stress; however, this factor is chosen very close to unity. Hence, the following relationships hold good:

$$Q = \frac{w_c}{w_w}$$

$$M_p = Kw_w$$

$$M_w = Kw_p$$

$$\frac{M_p}{M_w} = \frac{w_p}{w_w} = Q$$

$$\frac{M_p}{M_w} = \frac{Z_p\sigma_y}{Z_e\sigma_{all}}$$

$$\frac{M_p}{M_w} = \left(\frac{Z_p}{Z_e}\right).\frac{\sigma_y}{\sigma_{all}}$$

$$Q = (S)\frac{\sigma_y}{\sigma_{all}} \tag{2.19}$$

The above expression implies a simple straightforward relationship that connects the load factor and the factor of safety, as given below:

$$\text{Load factor} = (\text{shape factor})\ (\text{factor of safety})$$

Let us try to understand the above connectivity through a simple example. Consider a section under axial tension in working stress design. In axial tension, permissible (allowable) stress is $0.6\sigma_y$, which results in the following factor of safety:

$$\frac{\sigma_y}{\sigma_{all}} = \frac{\sigma_y}{0.66_y} = 1.66$$

For a rectangular section with a shape factor of 1.5, we get

$$\frac{M_p}{M_w} = Q = (1.5)(1.66) = 2.44$$

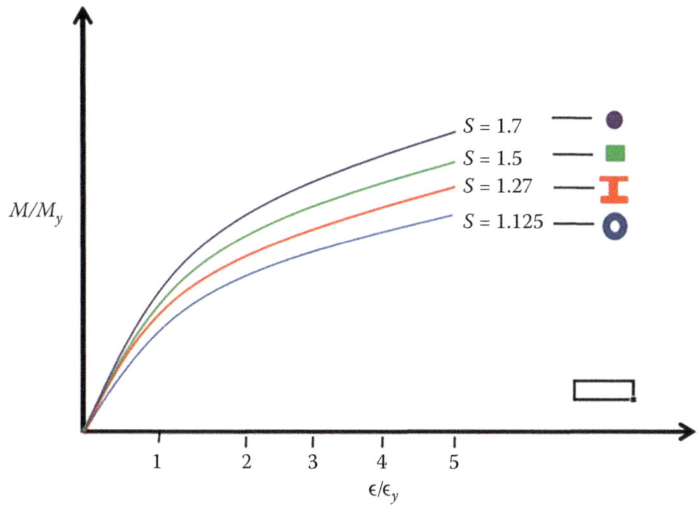

FIGURE 2.16 Shape factor for different cross section.

It is therefore clear from the above example that the plastic design also results in a good margin of safety, which is reflected through the load factor (Q). The shape factors for different geometries are shown in Figure 2.16.

2.10 ANALYSIS OF THE BEAM

When the structure is subjected to a system of loads and a sufficient number of plastic hinges are formed to transfer the moments, then segments of the beam between the plastic hinges are able to move without an increase in load. The system is now transformed into a *mechanism*. If an indeterminate structure has redundancy, r, then insertion of r plastic hinges makes it statically determinate. Any further hinge converts this statically determinate structure into a mechanism. Hence, for collapse, the number of plastic hinges required is $(r + 1)$. The following conditions are reached in a given structure:

Mechanism condition: The ultimate or collapse load is reached when a *mechanism* is formed. The number of plastic hinges developed should be just sufficient to form a *mechanism*.

Equilibrium condition: The following conditions hold good:

$$\Sigma F_x = 0$$

$$\Sigma F_y = 0$$

$$\Sigma M_{xy} = 0$$

Plastic moment condition: The bending moment at any section of the beam should not exceed the plastic moment capacity (M_p) of the section.

2.10.1 Collapse Mechanisms

When a system of loads acts on an elastic body, the body will show a resistance to deformation. However, if no resistance is set up against deformation in the body, then it is a *mechanism*. Various types of mechanisms enabling the prediction of possible failure modes of a structure are: (i) the beam mechanism, (ii) the sway or panel mechanism, (iii) the gable mechanism, and (iv) a combined mechanism. Figures 2.17 through 2.20 show the different mechanisms. Various combinations of independent

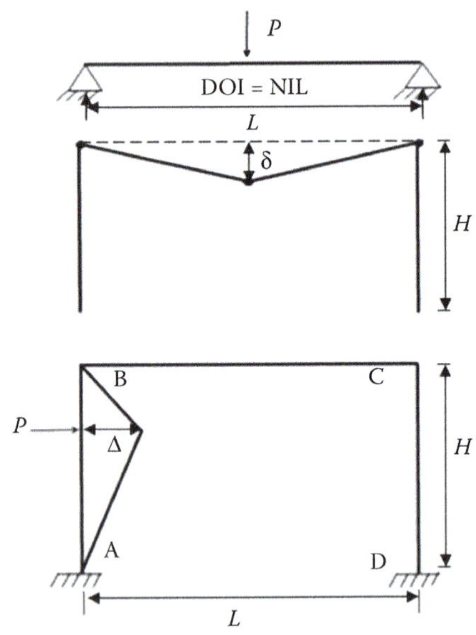

FIGURE 2.17 Beam mechanism of a frame.

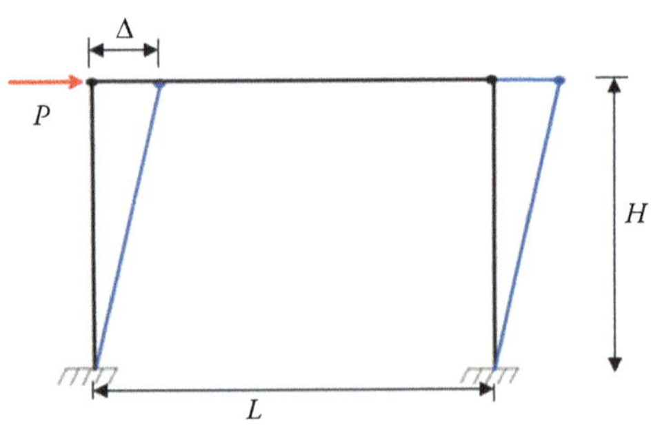

FIGURE 2.18 Sway mechanism of a frame.

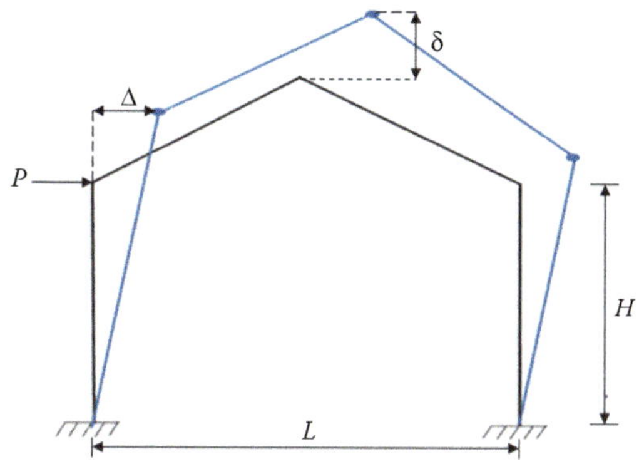

FIGURE 2.19 Gable mechanism of a frame.

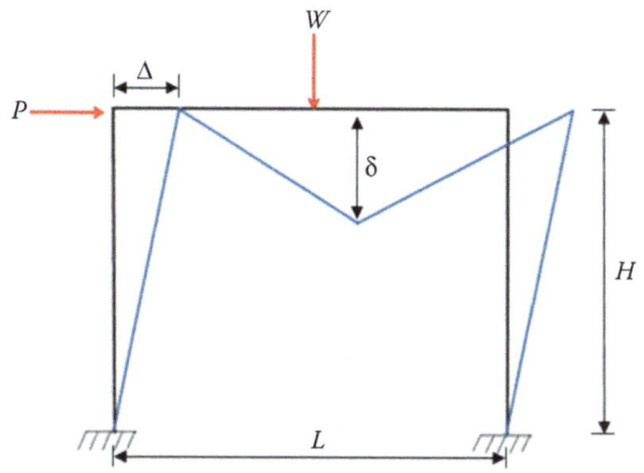

FIGURE 2.20 Combined mechanism of a frame.

mechanisms are assumed, depending upon whether the frame is made up of a strong-beam and weak-column combination, or vice versa. Failure is triggered by the formation of plastic hinges at critical sections all along the frame.

2.10.2 Static Theorem or Lower Bound Theorem

According to this theorem, for a given frame and load combination, if there exists any bending moment distribution throughout the frame, which is both safe and statically admissible with the set of loads Q, then the collapse load (W) will be less than or equal to the true collapse load (W_c). This infers a guideline statement that *the collapse load determined from the known distribution of bending moment will be either less than or equal to the true collapse load* W_c. The steps to solve a problem using the static theorem are as follows:

- Choose the set of redundant forces.
- Draw the free bending moment diagram, which is statically admissible and is caused by the set of redundant forces already chosen.

- Draw the redundant bending moment diagram, which arises from the support moments.
- Draw a combined bending moment distribution to determine the peak moments.
- Set all the peak moments to M_p.
- Check the number of plastic hinges required to form a *mechanism*.
- Determine the collapse load from equilibrium condition.
- Check whether the moment at any cross section exceeds M_p.
- The true collapse load will be the least of all the possible values of the collapse load thus obtained from the assumed bending moment distribution.

2.10.3 Kinematic Theorem or Upper Bound Theorem

According to this theorem, for a given frame and load combination, the value of collapse load corresponding to any assumed *mechanism* will be either equal to or greater than the true collapse load (W_c). This infers a guideline statement that *the collapse load determined from an assumed mechanism will be either greater than or equal to the true collapse load,* W_c. This method is iterative because the actual collapse mechanism for a known set of loading is not known in advance. As certain failure mechanisms need to be tried to arrive at the true collapse load, it is necessary to try various possible mechanisms, and hence is iterative. The steps to solve a problem using the kinematic theorem are as follows:

- Locate the points of possible plastic hinges for a given frame under the known set of loads.
- Determine the number of possible independent *mechanisms* and combined *mechanisms*. If the degree of indeterminacy of the frame is r, then the number of plastic hinges will not be greater than $(r + 1)$. The number of independent mechanisms will be $(N - r)$, where N is the number of possible plastic hinges.
- Solve the equation of the principle of virtual work to obtain the collapse load for each of the assumed mechanisms.
- The true collapse load will be the greatest of all the possible values of the collapse load thus obtained for all assumed mechanisms.

2.10.4 Principle of Virtual Work as Applied to Plastic Analysis

The principle of virtual work, as applied to plastic design is stated as follows:

Work done by external forces during small movements of the collapse mechanism is equal to the work absorbed by the plastic hinges.

2.10.5 Uniqueness Theorem or Combined Theorem

This theorem states that for a given frame and load combination, at least one safe and statically admissible bending moment distribution can be determined. This bending moment distribution is equal to the plastic moment of resistance (M_p), as sufficient plastic hinges will be formed in the critical sections to convert the frame into a

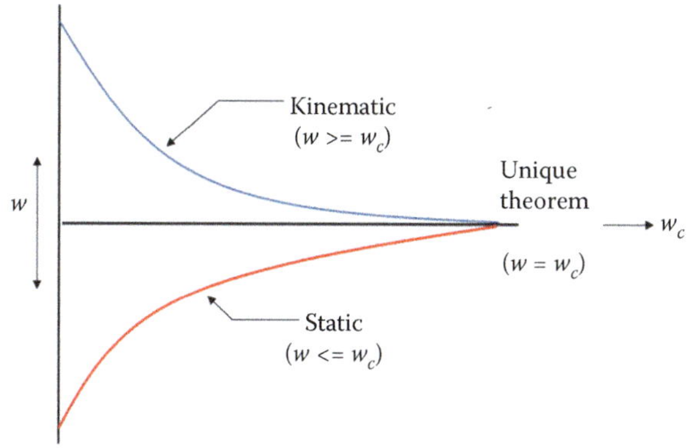

FIGURE 2.21 Uniqueness theorem.

mechanism. Then the corresponding load will be equal to the true collapse load (W_c). Figure 2.21 explains the uniqueness theorem.

2.11 EXAMPLE PROBLEMS FOR THE ESTIMATION OF THE TRUE COLLAPSE LOAD

2.11.1 Simply Supported Beam with a Central Concentrated Load

Let us consider a simply supported beam of span L, with a central concentrated load, W, as shown in the figure. As the beam is statically determinate, the degree of indeterminacy, $R = 0$. Hence, the number of plastic hinges required to form a *mechanism* is $(r + 1)$, which will be unity. The problem is solved using both the theorems for comparison.

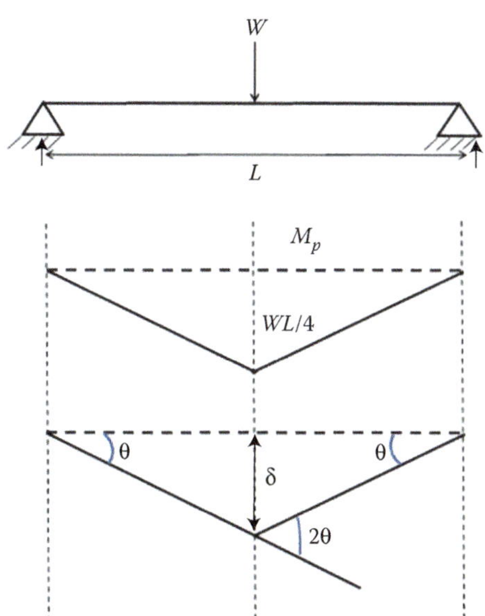

a. Using the static theorem:
Under the central concentrated load with simply supported ends, the maximum bending moment occurs at the mid-span (or at the section where the load is applied). A statically admissible bending moment distribution is drawn and the maximum bending moment value is assigned as M_p.

$$M_p = \frac{WL}{4}$$

$$W_c = \frac{4M_p}{L}$$

Hence, the collapse load will be the highest of the loads computed from the assumed (statically admissible) bending moment distribution, which is equal to W_c.

b. Using the kinematic theorem:
For the assumed beam mechanism, as shown in the figure, the principle of virtual work is applied to determine the collapse load.

$$\text{External virtual work} = w\delta$$

$$\text{Internal virtual work} = M_p(2\theta)$$

$$\tan\theta = \theta = \frac{\delta}{L/2} = \frac{2\delta}{L}$$

$$W\delta = 2M_p\left(\frac{2\delta}{L}\right)$$

$$W_c = \frac{4M_p}{L}$$

Hence, the collapse load will be the lowest of the loads computed from the assumed mechanism, which is equal to W_c. It can be readily seen that the collapse load determined using both the theorems converges to the same magnitude of W_c.

2.11.2 Simply Supported Beam with an Eccentric Load on the Span

Let us consider a simply supported beam of span L, with a concentrated load, W, as shown in the figure. As the beam is statically determinate, the degree of indeterminacy $R = 0$. Hence, number of plastic hinges required to form a *mechanism* is $(r + 1)$, which will be unity. The problem is solved using both the theorems for comparison.

a. Using the static theorem:

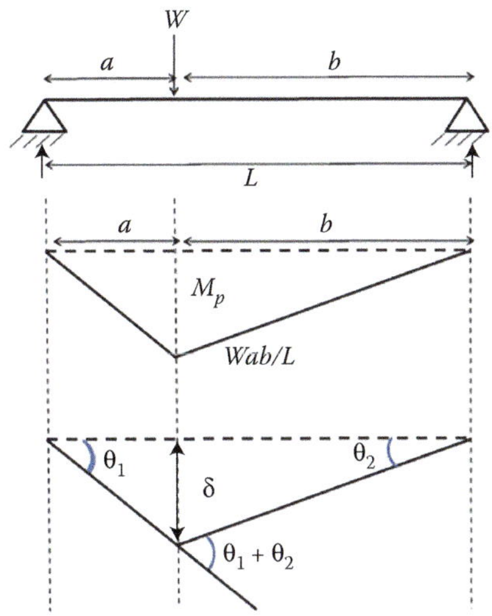

Under a concentrated load with simply supported ends, the maximum bending moment occurs at the section where the load is applied. A statically admissible bending moment distribution is drawn and the maximum bending moment value is assigned as M_p.

$$M_p = \frac{Wab}{L}$$

$$W_c = \frac{M_p L}{ab}$$

Hence, the collapse load will be the highest of the loads computed from the assumed (statically admissible) bending moment distribution, which is equal to W_c.

b. Using the kinematic theorem:
For the assumed beam mechanism, as shown in the figure, the principle of virtual work is applied to determine the collapse load.

$$W\delta = M_p(\theta_1 + \theta_2)$$

$$\theta_1 = \frac{\delta}{a};\ \delta = a\theta_1$$

$$\theta_2 = \frac{\delta}{b};\ \delta = a\theta_2$$

$$a\theta_1 = b\theta_2$$

$$\theta_1 = \left(\frac{b}{a}\right)\theta_2$$

$$W\delta = M_p\left[\frac{b}{a}(\theta_2 + \theta_2)\right]$$

$$= M_p\left(\frac{a+b}{a}\right)\theta_2$$

$$= M_p\frac{L}{2}(\theta_2)$$

$$W\delta = M_p\frac{L\delta}{ab}$$

$$W_c = \frac{M_pL}{ab}$$

Hence, the collapse load will be the lowest of the loads computed from the assumed mechanism, which is equal to W_c. It can be readily seen that the collapse load determined using both the theorems converges to the same magnitude of W_c.

2.11.3 Fixed Beam with a Central Concentrated Load

Let us consider a fixed beam of span L, with a concentrated load, W, as shown in the figure. As the beam is statically indeterminate, the degree of indeterminacy $R = 2$. Hence, the number of plastic hinges required to form a *mechanism* is $(r + 1)$, which will be three. The problem is solved using both theorems to show the comparison.

a. Using the static theorem:

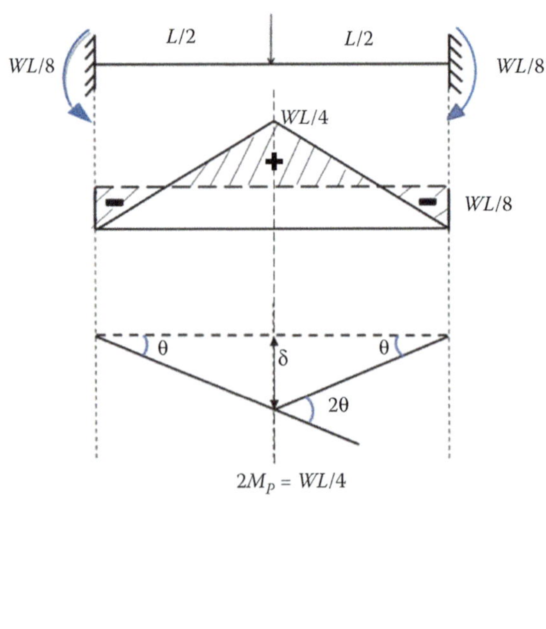

Under a concentrated load with fixed supports, the maximum bending moment occurs at the mid-span, where the load is applied. The statically admissible bending moment distribution is drawn and the maximum bending moment value is assigned as M_p.

$$2M_p = \frac{WL}{4}$$

$$W_c = \frac{8M_p}{L}$$

Hence, the collapse load will be the highest of the loads computed from the assumed (statically admissible) bending moment distribution, which is equal to W_c.

b. Using the kinematic theorem:
For the assumed beam mechanism, as shown in the figure, the principle of virtual work is applied to determine the collapse load.

$$W\delta = M_p(\theta + 2\theta + \theta)$$

$$= 4M_p\theta$$

$$\theta = \frac{8}{L/2} = \frac{2\delta}{L}$$

$$W\delta = 4M_p\left(\frac{2\delta}{L}\right)$$

$$W_c = \frac{8M_p}{L}$$

Hence, the collapse load will be the lowest of the loads computed from the assumed mechanism, which is equal to W_c. It can be readily seen that the collapse load determined using both the theorems converges to the same magnitude of W_c.

2.11.4 Fixed Beam with Uniformly Distributed Load

Let us consider a fixed beam of span L, with a uniformly distributed load for the complete span, as shown in the figure. As the beam is statically indeterminate, the degree of indeterminacy, $R = 2$. Hence, number of plastic hinges required to form a *mechanism* is $(r + 1)$, which will be three. The problem is solved using both the theorems for comparison.

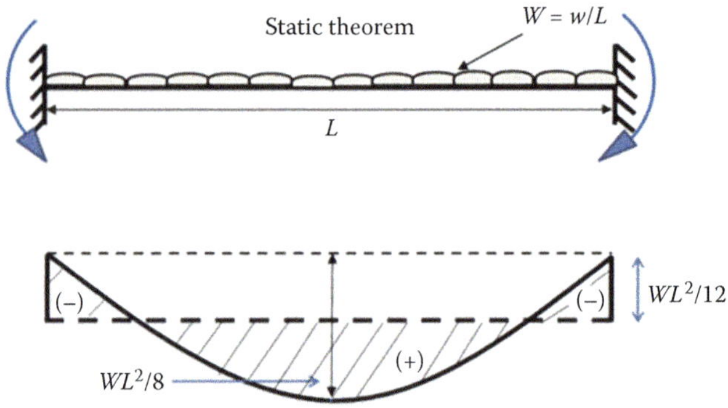

a. Using the static theorem:
   Under a uniformly distributed load with fixed supports, the maximum bending moment occurs at the mid-span, where the load is applied. The statically admissible bending moment distribution is drawn and the maximum bending moment value is assigned as M_p.

$$2M_p = \frac{wL^2}{8}$$

$$\frac{16M_p}{L} = (wL) = w_c$$

   Hence, the collapse load will be the highest of the loads computed from the assumed (statically admissible) bending moment distribution, which is equal to w_c.
b. Using the kinematic theorem:
   For the assumed beam mechanism, as shown in the figure, the principle of virtual work is applied to determine the collapse load.

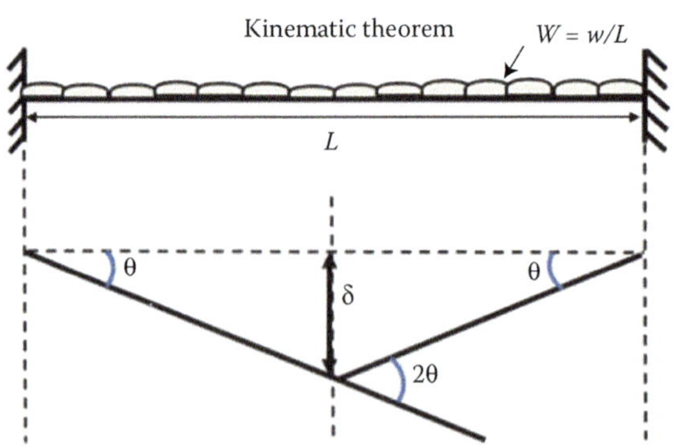

$$\frac{1}{2}L\delta w_c = M_p(\theta + 2\theta + \theta)$$

$$w_c L\delta = 4M_p\theta(2)$$

$$w_c\delta = 8M_p\theta$$

$$\theta = \frac{\delta}{L/2}$$

$$\theta = \frac{2\delta}{L}$$

$$w_c\delta = 8M_p\frac{2\delta}{L}$$

$$w_c = \frac{16M_p}{L}$$

Hence, the collapse load will be the lowest of the loads computed from the assumed mechanism, which is equal to w_c. It can be readily seen that the collapse load determined using both the theorems converges to the same magnitude.

2.11.5 Continuous Beam with Central Concentrated Load in Each Span

Let us consider a two-span continuous beam, each span L, with a central concentrated load, as shown in the figure. As the beam is statically indeterminate, the degree of indeterminacy is $R = 1$. Hence, the number of plastic hinges required to form a *mechanism* is $(r + 1)$, which will be two. The problem is solved using both the theorems for comparison.

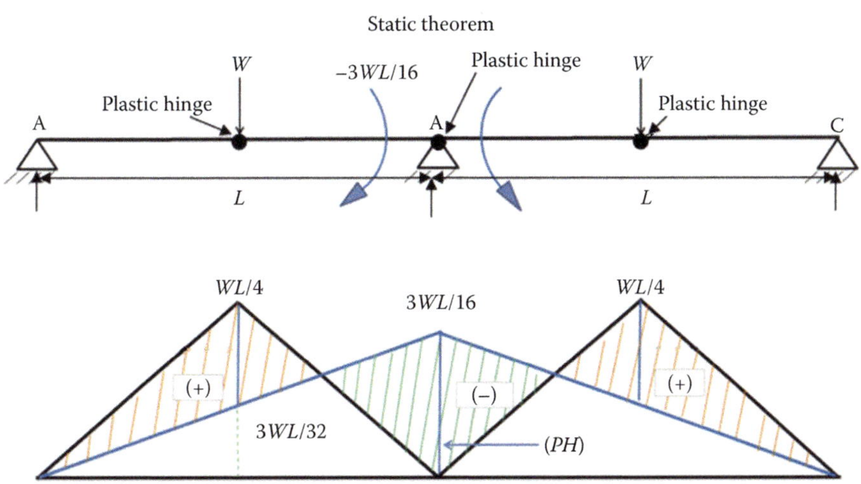

a. Using the static theorem:
   Under the central concentrated load on each span, the maximum bending moment occurs at the sections where the concentrated loads are applied. The statically admissible bending moment distribution is drawn and the maximum bending moment value is assigned as M_p.

$$\left(M_p + \frac{M_p}{2}\right) = \frac{wL}{4}$$

$$w_c = \frac{6M_p}{L}$$

   Hence, the collapse load will be the highest of the loads computed from the assumed (statically admissible) bending moment distribution, which is equal to w_c.

b. Using the kinematic theorem:
   For the assumed beam mechanism, as shown in the figure, the principle of virtual work is applied to determine the collapse load.

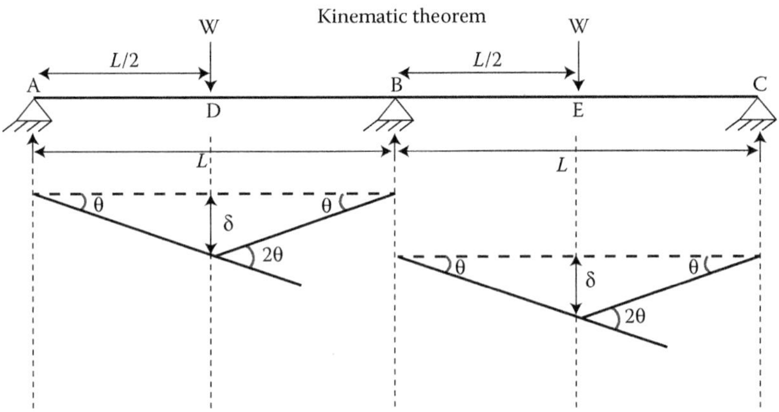

$$W\delta = M_p(3\theta)$$

$$\theta = \frac{\delta}{L/2}$$

$$= \frac{2\delta}{2}$$

$$w_c\delta = 3M_p\left(\frac{2\delta}{L}\right)$$

$$w_c = \frac{6M_p}{L}$$

Hence, the collapse load will be the lowest of the loads computed from the assumed mechanism, which is equal to w_c. It can be readily seen that the collapse load determined using both the theorems converges to the same magnitude.

2.11.6 Propped Cantilever with Uniformly Distributed Load

Let us consider a beam of span L, with one end fixed and the other end with a hinged support, as shown in the figure. The beam is subjected to a uniformly distributed load. As the beam is statically indeterminate, the degree of indeterminacy, $R = 1$. Hence, the number of plastic hinges required to form a *mechanism* is $(r + 1)$, which will be two. The problem is solved using both the theorems for comparison.

a. Using the static theorem:
   Under the uniformly distributed load, the maximum bending moment occurs at the section within the span, but it is not readily known. It is important to note that in this case, the section of maximum bending moment, which will also be the section where a plastic hinge will form, is not known. The statically admissible bending moment distribution is drawn and the maximum bending moment value is assigned as M_p. However, the distance of the section of maximum bending moment is still unknown.

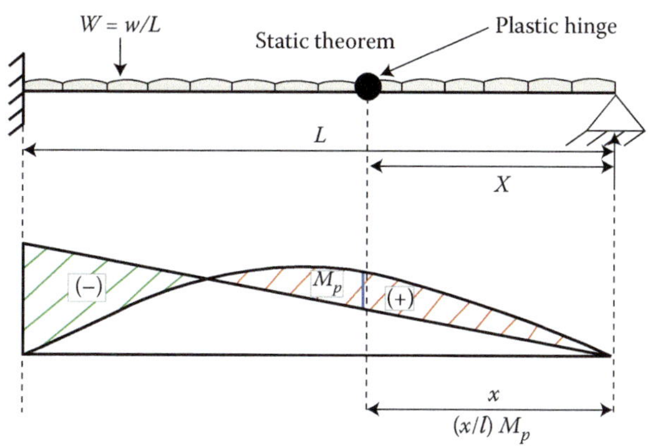

$$M_{xx} = \frac{w_c Lx}{2} - \frac{w_c x^2}{2}$$

$$= (M_p)x = \left[M_p + \frac{x}{L}M_p\right] = \frac{w_c Lx}{2} - \frac{w_c x^2}{2}$$

$$M_p\frac{(x - L)}{l} = \frac{w_c Lx}{2}$$

$$= \frac{w_c}{2}(L - x)$$

$$M_p = \frac{w_c}{2}\left[\frac{x(L - x)}{(L + x)}\right]$$

$$\frac{dM_p}{dx} = 0$$

$$= \frac{d}{dx}\left[\frac{w_c}{2}\left(Lx\frac{(L - x)}{(L + x)}\right)\right]$$

$$= \frac{w_c}{2}\left\{\frac{(L + x)(L - 2x) + x(L - x)}{(l + x)^2}\right\} = 0$$

$$x = 0.414L$$

$$M_p = 0.0857w_cL^2$$

$$W_c = 11.67\left(\frac{M_p}{L}\right)$$

Hence, the collapse load will be the highest of the loads computed from the assumed (statically admissible) bending moment distribution, which is equal to W_c.

b. Using the kinematic theorem:
For the assumed beam mechanism, as shown in the figure, the principle of virtual work is applied to determine the collapse load.

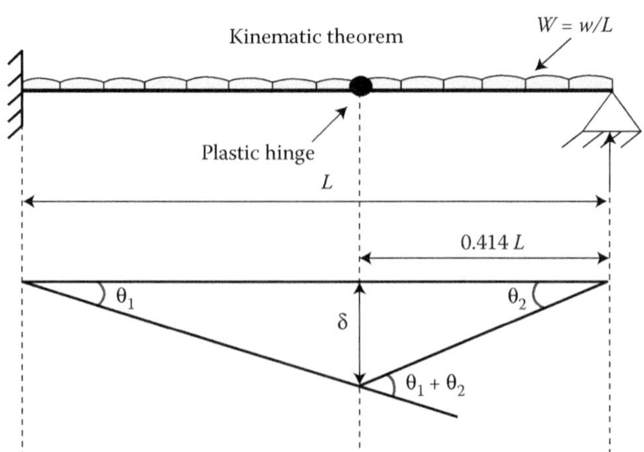

$$\frac{1}{2}(L\delta)w_c = M_p[\theta_1 + (\theta_1 + \theta_2)]$$

$$\frac{1}{2}w_c L\delta = M_p(2\theta_1 + \theta_2)$$

$$\theta_1 = \frac{\delta}{0.586L}$$

$$\theta_2 = \frac{\delta}{0.414L}$$

$$M_p = \left[\left(\frac{\delta}{0.586L}\right)\left(\frac{\delta}{0.414L}\right)\right]$$

$$= \left[\frac{2\delta}{0.586L} + \frac{\delta}{0.414L}\right]$$

$$w_c = \frac{11.66M_p}{L}$$

$$L^2 - 2Lx + xL - 2x^2 + L - x^2$$

$$-2Lx + Lx + Lx - 2x^2 - x^2 + L^2 = 0$$

$$-3x^2 + L^2 = 0$$

$$x^2 = \frac{L^2}{-3}$$

$$x = 0.414L$$

$$M_p = M_x$$

$$= \frac{w_c L}{2}\left[\frac{0.414(l - 0.414)}{(l + 0.414)}\right]$$

$$= \frac{W_C}{2}\left[\frac{0.414L - 0.1713}{0.414L}\right]$$

$$= W_C 0.0887$$

$$M_p = 0.0857 w_c L$$

$$w_c = 11.67 \frac{M_p}{L}$$

Hence, the collapse load will be the lowest of the loads computed from the assumed mechanism, which is equal to W_c. It can be readily seen that the collapse load determined using both the theorems converges to the same magnitude.

2.11.7 Single-Bay, Single-Story Frame

Let us consider a single-bay, single-story frame, as shown in the figure.

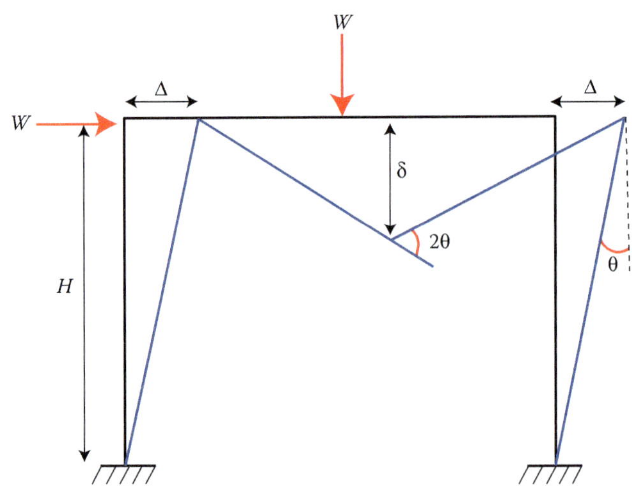

The frame is statically indeterminate to a third degree and, hence, four plastic hinges are required to form a *mechanism*. Possible locations of the plastic hinges could be the fixed supports, the beam–column junctions, or the point of concentrated load. The problem is solved using the kinematic theorem. Different independent mechanisms are assumed, and the collapse load is worked out for each assumed mechanism.

Beam mechanism

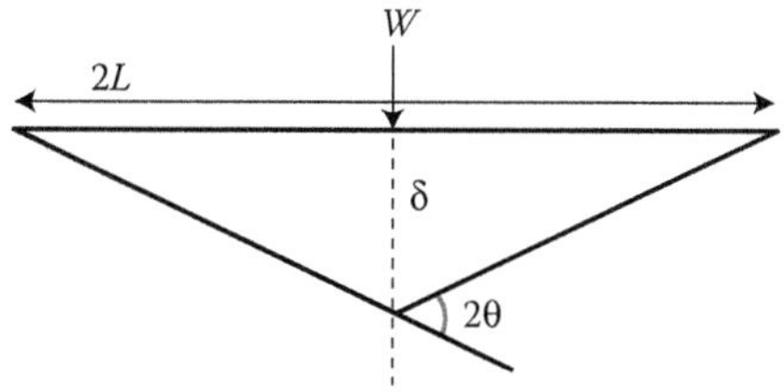

$$w\delta = M_p(4\theta)$$

$$= \frac{\delta}{L}$$

$$w_c\delta = 4M_p\frac{\delta}{L}$$

$$w_c = \frac{4M_p}{L}$$

Sway mechanism

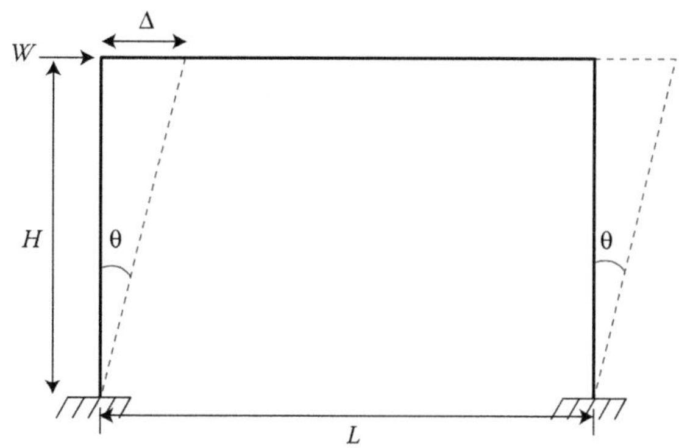

$$w_c\Delta = 4M_p\theta$$

$$w_c\Delta = \frac{\Delta}{L}$$

$$w_c\Delta = \frac{4M_p}{L}$$

Combined mechanism

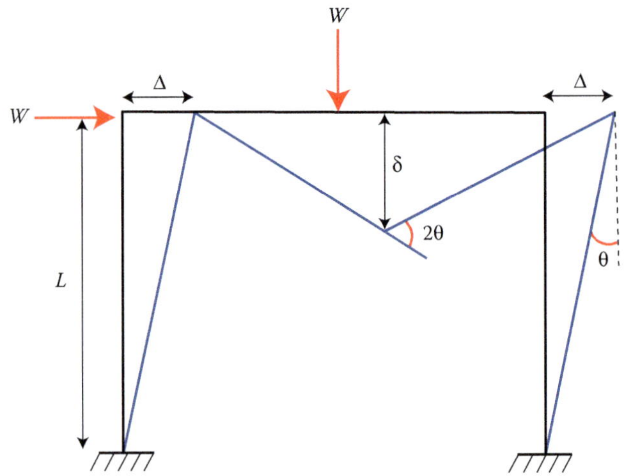

$$W(\Delta) + w\delta = M_p(\theta + 2\theta + 2\theta)$$

$$= 6M_p\theta$$

$$\theta = \frac{\Delta}{L} = \frac{\delta}{L}; \quad \Delta = \delta$$

$$2w_c\Delta = 6M_p\left(\frac{\Delta}{L}\right)$$

$$w_c = \frac{3M_p}{L}$$

Anticlockwise

$$M_p = M_p - \frac{w}{2}L$$

$$= M_p - \frac{3}{2}M_p = -\frac{M_p}{2} \ngtr M_p$$

Hence, the collapse load will be the lowest of the loads computed from the assumed mechanism, which is equal to $(3M_p/L)$.

2.11.8 Determining Yield Region

A simply supported beam of cross section 230 × 450 mm is fully loaded until the central section becomes fully plasticized. Demarcate the yield region. Take $\sigma_y = 250$ MPa = 250 N/m^2.

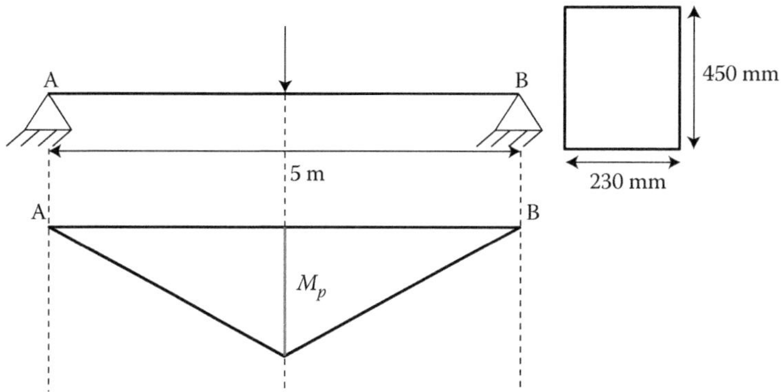

$$M_p = \sigma_y z_p$$

$$= \sigma_y (s) z_y$$

$$= (250)(1.5)\frac{bd^2}{12}$$

$$M_p = (250)(1.5)(230)\frac{450^2}{6} = 2.9 * 10^3$$

$$\frac{M}{M_p} = \left[1 - \frac{e^2}{3h^2}\right]$$

$$M_{1.62} = [1.95 * 10^3]\ \text{kNm}$$

$$M_{1.8} = [2.088 * 10^3]\ \text{kNm}$$

$$M_{1.9} = [2.204 * 10^3]\ \text{kNm}$$

$$M_{2.0} = [2.32 * 10^3]\ \text{kNm}$$

$$M_{2.1} = [2.436 * 10^3]\ \text{kNm}$$

$$M_{2.2} = [2.552 * 10^3]\text{ kNm}$$

$$M_{2.3} = [2.668 * 10^3]\text{ kNm}$$

$$M_{2.4} = [2.784 * 10^3]\text{ kNm}$$

$$M_{2.8} = [2.9 * 10^3]\text{ kNm}$$

To find the depth of the elastic core, the following relationship is used:

$$\left(1 - \frac{M}{M_p}\right)3h^2 = e^2$$

$$\left(1 - \frac{M_{1.67}}{M_p}\right)3 * 4^2 = e_{1.67}$$

$$3h^2\left(1 - \frac{M_{1.8}}{M_p}\right) = e_{1.8}$$

Typical values of the depth of the elastic core can be determined using similar relationships, as above. The yield region can then be demarcated.

2.12 FAILURE THEORIES

In practice, engineering materials are observed to fail in two ways: (i) by yielding and (ii) by fracture. *Yielding* is a permanent deformation, which occurs due to the pronounced sliding of planes through crystalline grains of the material. It takes place without any actual rupture of the material. The functional utility of the member will be lost after a particular amount of yielding takes place. Therefore, for all practical purposes, yielding is considered a criteria of failure. *Fracture* is failure in which separation occurs at a cross section normal to the direction of tensile stresses. While this fracture criterion is applicable to brittle materials, in practice a limit of about 5% elongation is taken as a partition line between ductile and brittle materials. It is therefore important to understand the yield strength and the limitations associated with it in order to decide the type of failure. For structural members subjected to a uniaxial stress system, limiting allowable stress for design is usually obtained from the tensile test of the specimen. While there are a few problems associated with the simple tensile test, it is important to know that the structures are subjected to a complex stress state. Many factors influence their failure, including: (i) state of stress, (ii) type of load, and (iii) heat treatment process. It is imperative to establish the criteria for the behavior of material/members under the combined state of stress in order to obtain the yield strength, as yield strength in turn governs the failure load.

2.12.1 Limitations Associated with Simple Tensile Test

In a simple tension test, when the specimen starts yielding, the following quantities are also reached simultaneously:

- Principal stress (σ_{max}) reaches the yield point stress (σ_{yp} or σ_{ult}) of the material.
- Maximum shear stress reaches the yield point stress, as given below:

$$\tau_{max} = 0.5\,\sigma_{max}$$

$$\tau_{yp} = 0.5\,\sigma_{yp}$$

- Tensile strain (ϵ) reaches the yield point strain (ϵ_{yp}).
- Total strain energy (U), absorbed by unit volume of the material also reaches the ultimate value, as given below:

$$U_{yp} = \left(\frac{\sigma_{yp}^2}{2E}\right)$$

- Strain energy distribution (U_d), absorbed per unit volume of the material, reaches the yield point value, as given below:

$$(U_d)_{yp} = \frac{1+\mu}{3E}\sigma_{yp}^2$$

- Octahedral shear stress reaches the yield point value, as given below:

$$\tau_{0yp} = \frac{\sqrt{2}}{3}\sigma_{yp} = 0.47\sigma_{yp}$$

In a multiaxial stress state, the above values will not be attained simultaneously. It is therefore important to choose any one of the above quantities to compute the ultimate load that will not cause inelastic deformation of the material. Different failure theories are discussed below to highlight the failure stress as projected by each one of them.

2.12.2 Maximum Principal Theory

This theory is based on Rankin's theory. It assumes that failure or fracture of the material occurs when the maximum principal stress at any section attains the critical value, regardless of other stresses. The critical value of the stress (σ_{ult}) will be determined by a simple tensile test. Failure is defined either by large or excessive elongation, or by fracture.

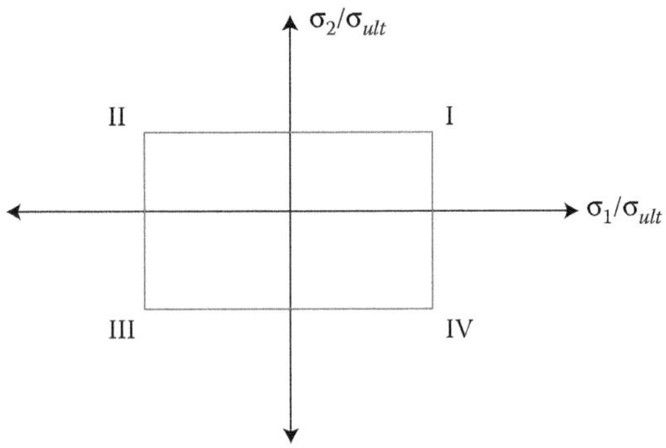

For a complex stress state, the maximum principal stress is given by

$$\sigma_1 = \frac{\sigma_x + \sigma_y}{2} \pm \frac{1}{2}\sqrt{(\sigma_x \sigma_y)^2 + z^2} \tag{2.20}$$
$$= \sigma_{ult} \text{(obtained from the simple tensile test)}$$

The figure shows the plot of the stress state using this theory. The sides of the square are given by

$$\frac{\sigma_1}{\sigma_{ult}} = \pm 1 \tag{2.21}$$

$$\frac{\sigma_2}{\sigma_{ult}} = \pm 1 \tag{2.22}$$

It can be seen that failure occurs when the point lies on the periphery of the stress plot or outside it. Experimental studies indicate that this theory shows good agreement for brittle materials in all ranges, and stress values for both the stresses in a biaxial stress state are tensile in nature. Failure in such cases is by fracture.

2.12.3 Maximum Shear Stress Theory (Tresca's)

This theory was proposed by Guest and Tresca, based on observations made on ductile materials. This theory assumes that slipping occurs by yielding along the critically oriented planes. The main features are as follows:

- Maximum stress alone produces inelastic deformation.
- If $\sigma_1 = \sigma_2$, then there will be no influence on the inelastic deformation.
- Failure will occur when the maximum shear stress reaches the value of that of the simple tensile test for yield point.

For a biaxial stress state, the stress plot is shown in the figure.

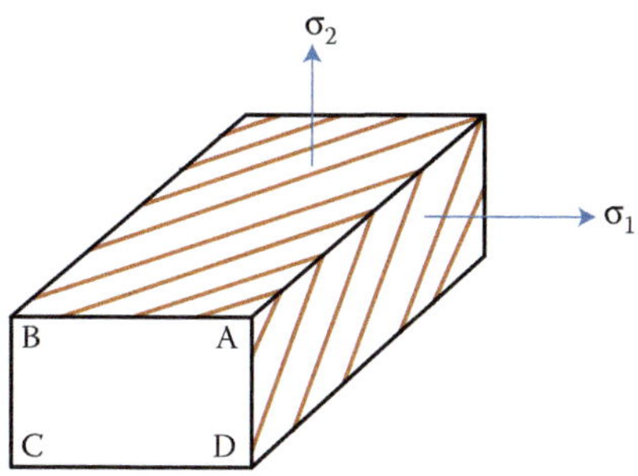

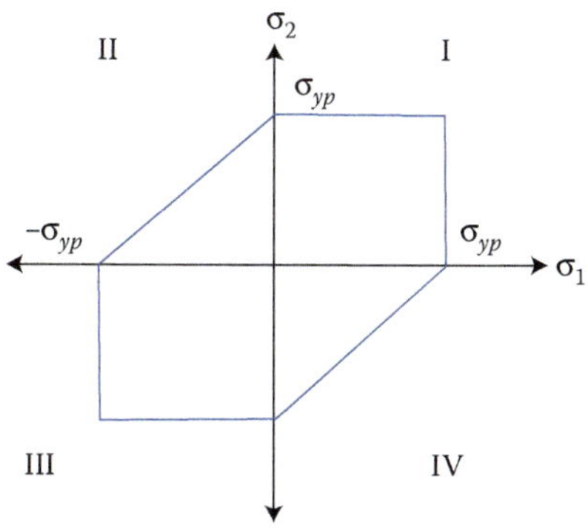

$$\sigma_1 > \sigma_2 > \sigma_3$$

$$\tau_{max} = \frac{\sigma_1 - \sigma_2}{2} = \tau_{yp} = \frac{\sigma_{yp}}{2} \tag{2.23}$$

$$\sigma_1 - \sigma_2 = \sigma_{yp}$$

$$-(\sigma_1 - \sigma_2) = \sigma_{yp}$$

$$\sigma_1 - \sigma_2 = \pm\sigma_{yp}$$

$$\frac{\sigma_1}{\sigma_{yp}} - \frac{\sigma_2}{\sigma_{yp}} = \pm 1 \tag{2.24}$$

$$\sigma_{2=0,}\sigma_{1=\pm}\sigma_{yp}$$

$$\sigma_{1=0,}\sigma_{2=\pm}\sigma_{yp}$$

If a point having stress values (σ_1, σ_2) lies inside the hexagon in the figure, it is presumed that no yielding of the material has occurred. When the stress value falls on the periphery of the hexagon, then the material is assumed to have undergone inelastic deformation.

Similarly, for a triaxial stress state, we can write the following statements:

1. If $\sigma_1 = \sigma_2 = \sigma_3$, then shear stress is very small.
2. Failure will be by fracture and not by yielding.

It is better to use the maximum principal stress theory in this case to compute the failure load. It can also be seen that the maximum shear stress theory is applicable to ductile materials for the state of stress in which comparatively large shear stresses are developed. For a tension test, in the case of pure shear, where maximum shear stress is developed, the shear elastic limit of a ductile material will be given by

$$\tau_{yp} = \frac{\sigma_{yp}}{2} = 0.5\sigma_{yp} \tag{2.25}$$

Experiments show that in pure shear, this value is $0.57\sigma_{yp}$. Hence, in such cases, maximum shear stress theory gives conservative results.

2.12.4 Maximum Strain Theory (St. Venant's)

In a given complex stress state, yielding at a point in any material begins when the maximum strain in the material exceeds the strain corresponding to the yield point.

If σ_1 and σ_2 are principal stresses such that $(\sigma_1 > \sigma_2)$, then strain in the direction of σ_1 will be given by

$$\varepsilon_1 = \frac{\sigma_1}{E} - \frac{\mu\sigma_2}{E} \not> \frac{\sigma_{yp}}{E} \tag{2.26}$$

$$\sigma_1 - \mu\sigma_2 = \sigma_{yp} \tag{2.27}$$

If σ_1 and σ_2 are both tensile, then σ_1 can even be greater than σ_{yp}. But if σ_2 is compressive in nature, then $\sigma_1 < \sigma_{yp}$.

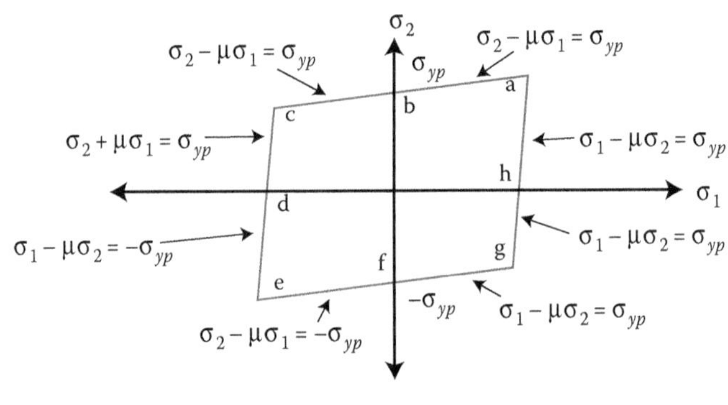

This implies that when both the stresses in a biaxial stress state are tensile, then $\sigma_1 > \sigma_{yp}$ even without causing yielding in the material. Hence, the maximum strain theory is an improvement on the principal stress theory. However, it is important to note that this theory does not give satisfactory results in ductile materials and is more suitable for brittle materials. A graphical representation of the stress state is shown.

2.12.5 Total Strain Energy Theory

This theory was proposed by Haigh. It states that inelastic action or yielding at a point begins when the energy/unit volume absorbed at a point is equal to the energy under the uniaxial stress state, as computed from a simple tensile test. In this case, failure does not depend on the state of stress, but is governed by the energy stored in the material/unit volume.

For a given triaxial stress, where $\sigma_1 > \sigma_2 > \sigma_3$

$$\varepsilon_1 = \frac{1}{E}[\sigma_1 - \mu(\sigma_2 + \sigma_3)]$$

$$\varepsilon_2 = \frac{1}{E}[\sigma_2 - \mu(\sigma_3 + \sigma_1)]$$

$$\varepsilon_3 = \frac{1}{E}[\sigma_3 - \mu(\sigma_1 + \sigma_3)]$$

Strain energy per unit volume is given by

$$U = \frac{1}{2}\sigma_1\varepsilon_1 + \frac{1}{2}\sigma_2\varepsilon_2 + \frac{1}{2}\sigma_3\varepsilon_3 \tag{2.28}$$

Substituting strain, in terms of stress, in the above equation, we get

$$U = \frac{1}{2}\sigma_1\left[\frac{1}{E}(\sigma_1 - \mu(\sigma_2 + \sigma_3))\right] + \frac{1}{2}\sigma_2\left[\frac{1}{E}(\sigma_2 - \mu(\sigma_3 - \sigma_1))\right] + \frac{1}{E}\sigma_3\left[\frac{1}{E}(\sigma_3 - \mu(\sigma_1 + \sigma_3))\right]$$

$$U = \frac{1}{2E}[\{\sigma_1^2 - \mu(\sigma_1\sigma_2 + \sigma_2\sigma_3)\} + \{\sigma_2^2 - \mu(\sigma_2\sigma_3 + \sigma_1\sigma_2)\} + \{\sigma_3^2 - (\sigma_3\sigma_1 + \sigma_3\sigma_2)\}]$$

$$U = \frac{1}{2E}[\sigma_1^2\sigma_2^2\sigma_3^2 - 2\mu(\sigma_1\sigma_2 + \sigma_2\sigma_3 + \sigma_3\sigma_1)] \tag{2.29}$$

For a biaxial stress state, the following relationship holds good:

$$U = \frac{1}{2E}[\sigma_1^2 + \sigma_2^2 - 2\mu(\sigma_1\sigma_2)] \tag{2.30}$$

For a uniaxial stress state, $\sigma_2 = \sigma_3 = 0$. Substituting in the above, we get

$$U = \frac{1}{2E}\sigma_1^2 = \frac{\sigma_{yp}^2}{2E} \tag{2.31}$$

Hence, for failure by the total strain energy theory, the following relationships hold good:

For a triaxial stress state, we get

$$\sigma_1^2 + \sigma_2^2 + \sigma_3^2 - 2\mu(\sigma_1\sigma_2 + \sigma_2\sigma_3 + \sigma_3\sigma_1) = \sigma_{yp}^2 \tag{2.32}$$

For a biaxial stress state, we get

$$\sigma_1^2 + \sigma_2^2 - 2\mu\sigma_1\sigma_2 = \sigma_{yp}^2 \tag{2.33}$$

The above equation represents an ellipse with major and minor axes inclined at 45° to both the stress axes, as shown in the figure.

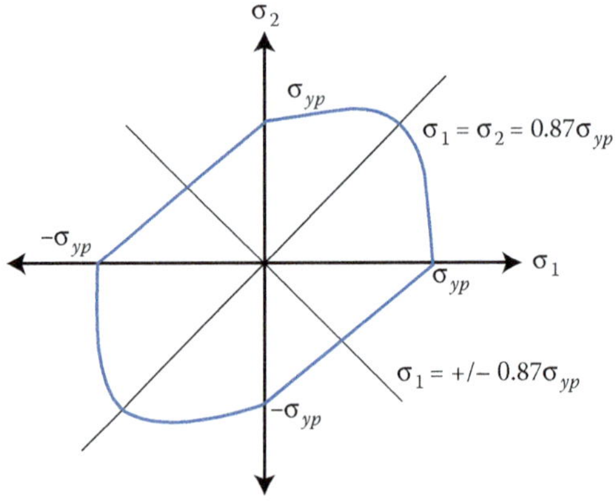

$$\text{for } \sigma_2 = 0, \sigma_1^2 = \sigma_{yp}$$

$$\sigma_1 = 0, \sigma_2 = \pm\sigma_{yp}$$

$\sigma_1 = \sigma_2 = \sigma$, stress of same intensity, we get

$$\sigma = \pm\frac{\sigma_{yp}}{\sqrt{2(1-\mu)}} \tag{2.34}$$

For both stresses tensile and Poisson ratio 0.33, we get

$$\sigma = \pm 0.87\sigma_{yp} \tag{2.35}$$

for $\sigma_1 = \sigma$, $\sigma_2 = -\sigma$, we get

$$\begin{aligned} \sigma^2 + \sigma^2 &= 2\mu\sigma^2 = \sigma_{yp}^2 \\ \sigma &= \frac{\sigma_{yp}}{\sqrt{2(1+\mu)}} \\ &= \frac{\sigma_{yp}}{\sqrt{2(1+(1/3))}} = 0.61\sigma_{yp} \end{aligned} \tag{2.36}$$

2.12.6 Maximum Distortion Theory (von Mises Theory)

This theory of failure was established by von Mises and Hencky. It states that inelastic action at any point in a material will occur only when the strain energy per unit volume, absorbed by the material, corresponds to the yield point stress, as obtained by a simple tensile test.

For a triaxial stress state, total strain energy per unit volume is given by

$$U = \frac{1}{2E}[\sigma_1^2 + \sigma_2^2 + \sigma_3^2 - 2\mu(\sigma_1\sigma_2 + \sigma_2\sigma_3 + \sigma_3\sigma_1)] \tag{2.37}$$

Neglecting the higher powers, the volumetric strain can be expressed as

$$\begin{aligned} \varepsilon_v &= \sigma_1 + \sigma_2 + \sigma_3 \\ \varepsilon_1 &= \frac{1}{E}[\sigma_1 - \mu(\sigma_2 + \sigma_3)] \\ \varepsilon_2 &= \frac{1}{E}[\sigma_2 - \mu(\sigma_3 + \sigma_1)] \\ \varepsilon_3 &= \frac{1}{E}[\sigma_3 - \mu(\sigma_1 + \sigma_2)] \\ \varepsilon_v &= \frac{1}{E}[\sigma_1 + \sigma_2 + \sigma_3 - 2\mu(\sigma_1 + \sigma_2 + \sigma_3)] \\ &= \frac{1-2\mu}{E}[\sigma_1 + \sigma_2 + \sigma_3] \end{aligned} \tag{2.38}$$

From the above equation, we can infer that the volumetric strain is proportional to the sum of three principal stresses. When the summation vanishes, then the volumetric strain vanishes and the body will be subjected to pure distortion. Total strain energy consists of two parts: (i) that associated with the volumetric change of the body and (ii) that associated with change in shape or distortion.

For no volumetric change in the body, the summation of principal stresses should be equated to *zero*. We get

$$\sigma_1 + \sigma_2 + \sigma_3 = 0$$

If $\sigma_1 = \sigma_2 = \sigma_3 = P$, then there will be no distortion in the body. Hence, the following relationship holds good:

$$P = \frac{\sigma_1 + \sigma_2 + \sigma_3}{3}$$

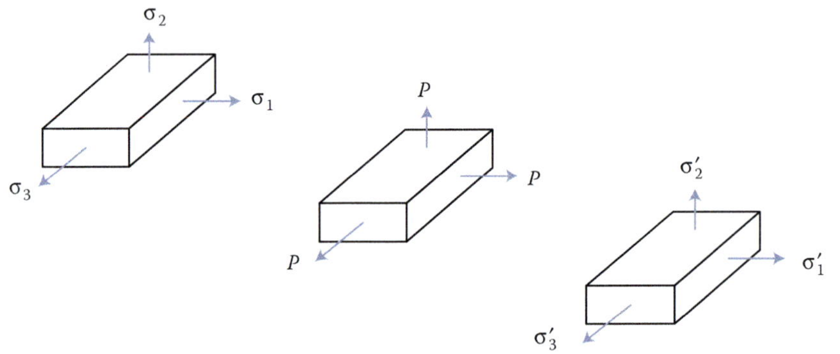

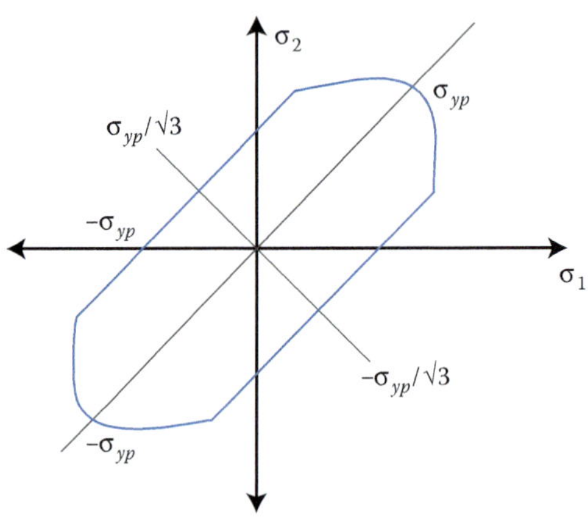

$$\begin{aligned}\sigma_1 &= P + \sigma_1' \\ \sigma_2 &= P + \sigma_2' \\ \sigma_3 &= P + \sigma_3'\end{aligned}$$

Adding, we get

$$\sigma_1 + \sigma_2 + \sigma_3 = 3P + \sigma^1 + \sigma_2^1 + \sigma_3^1$$

For no distortion, the following relationship holds good:

$$P = 1/3\{\sigma_1 + \sigma_2 + \sigma_3\}$$

and hence for $\sigma_1^1 + \sigma_2^1 + \sigma_3^1 = 0$, this condition can cause pure distortion with no change in volume.

Let $\sigma_1 = \sigma_2 = \sigma_3 = P$ be substituted in the following equation. We get

$$U = \frac{1}{2E}[\sigma_1^2 + \sigma_2^2 + \sigma_3^2 - 2\mu(\sigma_1\sigma_2 + \sigma_2\sigma_3 + \sigma_3\sigma_1)] \tag{2.39}$$

$$U_v = \frac{3P^2}{2E}(1 - 2\mu) \tag{2.40}$$

$$P = \frac{(\sigma_1 + \sigma_2 + \sigma_3)}{3}$$

Substituting, we get

$$U_v = \left(\frac{1 - 2\mu}{6E}\right)(\sigma_1 + \sigma_2 + \sigma_3)^2 \tag{2.41}$$

Distortion energy may be obtained by subtracting the above from the total strain energy. We get

$$U_{distortion} = U - U_v$$

$$U_{distortion} = \frac{1 + \mu}{6E}[(\sigma_1 - \sigma_2)^2 + (\sigma_2 - \sigma_3)^2 + (\sigma_3 - \sigma_1)^2] \tag{2.42}$$

Distortion energy for material under a uniaxial stress state in a simple tensile test is given by

$$\sigma_2 = \sigma_3 = 0$$

$$\sigma_1 = \sigma_{yp}$$

$$u_{distortion} = \frac{1 + \mu}{3E}\sigma_{yp}^2 \tag{2.43}$$

The condition of yielding, according to the maximum distortion theory, is given by

$$\begin{aligned}(\sigma_1 - \sigma_2)^2 + (\sigma_2 - \sigma_3)^2 + (\sigma_3 - \sigma_1)^2 &= 2\sigma_{yp}^2 \\ (\sigma_1^2 + \sigma_2^2 - 2\sigma_1\sigma_2) + \sigma_2^2 + \sigma_1^2 &= 2\sigma_{yp}^2\end{aligned} \tag{2.44}$$

For a plane stress problem, σ_3 is set to zero. Then we get
$\sigma_1^2 + \sigma_2^2 - 2\sigma_1\sigma_2 = \sigma_{yp}^2$ which is an equation of ellipse.
Maximum distortion theory has good agreement with the experimental results for ductile materials.

2.12.7 Comparison of Failure Theories

Based on the above discussions, it can be seen that the maximum principal stress theory appears to be the best applicable failure theory for brittle materials, but is found to be unsafe for ductile materials. If one of the principal stresses at any point is very large in magnitude in comparison with the others, then all theories will yield almost the same results. Discrepancies of the failure theories are greatest for stress values that are dissimilar in nature. For example, if a body is subjected to biaxial stress state with both the stresses of similar nature, then the yield strength predicted by all the above theories does not vary much; variation is significant only when the stresses are dissimilar in nature. A comparative plot is shown.

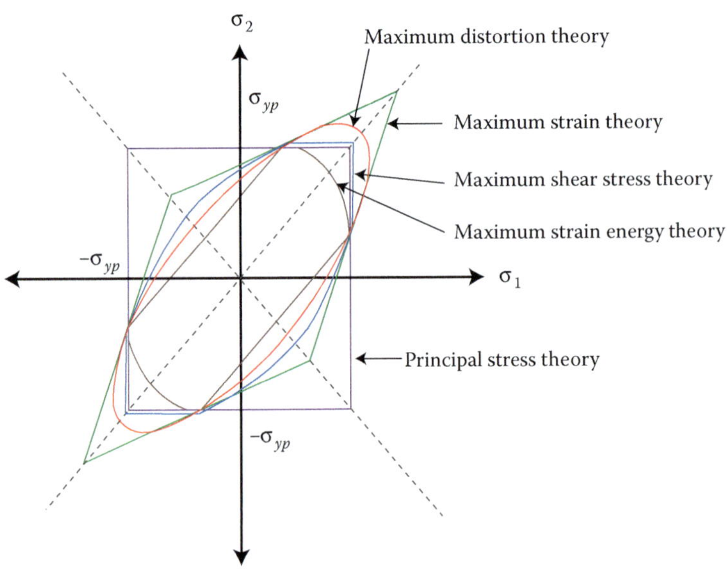

The figure shows that, for a biaxial state, ($\sigma_3 = 0$) and the material has the same tensile strength. The following discussions are interesting and important:

a. By comparing St. Venant's theory with the maximum principal stress theory, it is seen that tension causes yielding at higher values of yield strength.

This is due to the fact that since tension in one direction reduces strain in the normal direction, stresses of same magnitude and nature will cause yielding at higher stress values. For similar stresses (e.g., both are tensile), the following relationship holds good:

$$\sigma_1 - \mu\sigma_2 = \sigma_{yp}$$

For stresses with equal magnitude

$$\sigma_1 = \sigma_2 = \left\{ \frac{\sigma_{yp}}{(1-\mu)} \right\}$$

b. In the case of dissimilar stresses, maximum strain theory indicates yielding at the stress value $\{\sigma_{yp}/(1-\mu)\}$.
c. An irregular hexagon shows that the maximum stress theory (Tresca's) coincides with that of the values of principal stress theory, wherever the stresses are of the same nature. For stresses unalike in nature, the difference between these two theories is significant.
d. Maximum shear stress theory has only a small deviation from the maximum distortion theory in the 2nd and 4th quadrants.

2.12.8 Example Problems in Theories of Failure

a. The maximum principal stress of a member is 200 N/mm^2 (tensile) and the minor stress is not known (compressive). If the yield strength of the material is 300 N\mm^2 (same in tension and compression), find the minor principal stress using different theories of failure. Take $\mu = 0.25$.
   1. Maximum stress theory
   2. Maximum shear stress theory
   3. Total shear stress theory
   4. Maximum distortion theory

Solution

For maximum stress theory

$$\sigma_1 - \mu\sigma_2 = \sigma_{yL}$$

$$200 + 0.25\sigma_2 = 300$$

$$\sigma_2 = 400 \text{ N/mm}^2$$

For maximum shear stress theory

$$\sigma_1 - \sigma_2 = \sigma_{yp}$$

$$200 - \sigma_2 = 300$$

$$\sigma_2 = 100 \text{ N/mm}^2$$

For total shear stress theory

$$\sigma_1^2 + \sigma_2^2 - \mu\sigma_1\sigma_2 = \sigma_{yp}^2$$

$$\sigma_1^2 + \sigma_2^2 + \mu\sigma_1\sigma_2 = \sigma_{yp}^2$$

$$200^2 + \sigma_2^2 + (2 * 0.25)(200)\sigma_2 = 300^2$$

$$\sigma_2 = 179.13 \text{N/mm}^2$$

For maximum distortion theory

$$\sigma_1^2 + \sigma_2^2 - \sigma_1\sigma_2 = \sigma_{yp}^2$$

$$200^2 + \sigma_2^2 + (200)\sigma_2 = 300^2$$

$$\sigma_2 = 144.98 \text{ N/mm}^2$$

b. Compare the permissible diameter of a shaft shown in the figure subjected to torsion ($\mu = 0.3$)

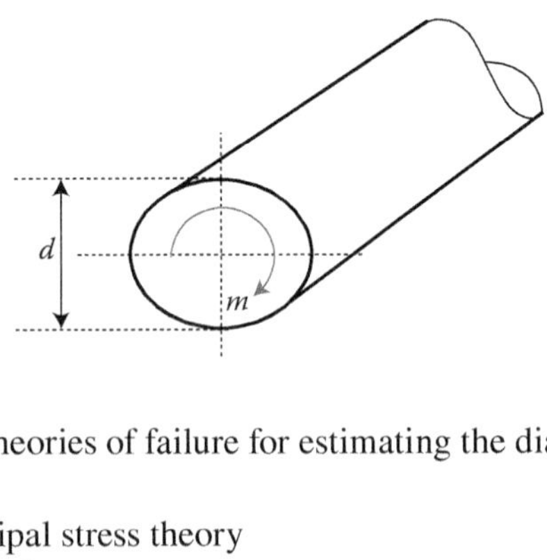

Use the following theories of failure for estimating the diameter:

1. Maximum principal stress theory
2. Maximum strain theory
3. Maximum shear stress theory
4. Maximum strain energy theory

Solution

$$(\sigma_{yp})_T = (\sigma_{yp})_c$$

According to the failure theories, conditions for yielding are as follows:

Maximum principal stress theory = $\sigma_1 = \sigma_{yp}$
Maximum strain theory = $\sigma_1 - \mu\sigma_2 = \sigma_{yp}$
Maximum shear stress theory = $\sigma_1 - \sigma_2 = \sigma_{yp}$
Maximum strain energy theory = $\sigma_1^2 + \sigma_2^2 - 2\mu\sigma_1\sigma_2 = \sigma_{yp}^2$

For the pure shear case, the following relationship holds good:

$$\sigma_1 = \sigma_2 = \tau$$

Hence, the above equations can be rewritten as

$$\tau_{yp} = \sigma_{yp}$$

$$\tau_{yp} = \frac{\sigma_{yp}}{(1+\mu)}$$

$$\tau_{yp} = \frac{\sigma_{yp}}{2}$$

$$\tau_{yp} = \frac{\sigma_{yp}}{\sqrt{2(1+\mu)}}$$

For the design of a circular shaft in tension, the permissible working stress in shear is given by

$$\tau_w = \frac{\tau_{yp}}{FOS}$$

For torsional moment, $\tau_w = 16M_{tor}/\pi d^3$
For the maximum principal stress theory, we get

$$\frac{\tau_{yp}}{FOS} = \frac{16M_t}{\pi d_1^3} = \frac{\sigma_{yp}}{FOS}$$

For the maximum strain theory, we get

$$\frac{16M_t}{\pi d_2^3} = \frac{0.76\sigma_{yp}}{FOS}$$

By comparing, we get

$$\frac{d_1}{d_2} = 0.913 \quad \text{or} \quad d_1 : d_2 = 1 : 1.095$$

For the maximum shear stress theory, we get

$$\frac{16M_t}{\pi d_3^3} = \frac{0.5\sigma_{yp}}{FOS}$$

By comparing, we get

$$d_1 : d_3 = 1 : 1.26$$

For the total strain energy theory, we get

$$\frac{16M_t}{\pi d_4^3} = \frac{0.62\sigma_{yp}}{FOS}$$

By comparing, we get

$$d_1 : d_4 = 1 : 1.17$$

Now for the design, we get

$$d_1 : d_2 : d_3 : d_4 = 1 : 1.09 : 1.26 : 1.17$$

2.13 ULTIMATE CAPACITY OF TUBULAR JOINTS

In marine structures, members are mostly fabricated using thin-walled tubular section (e.g., legs and pontoons are tubular sections). Generally, these members are subjected to impact forces in addition to the combined action of bending, axial, and lateral forces. In general, joints are the prospective location of plastic hinges. They are subjected to the combined action of membrane forces and moments. Hence, these joints need a thorough examination for their load-carrying capacity. These joints have high reserve strength and capability to redistribute the stresses. Marine structures are subjected to different types of loads, namely, gravity loads, wind loads, wave loads, and current loads. Under the above loads and their combination, normal stresses in the bracings and legs are obtained for design purposes. Different levels of analyses of tubular joints are carried out:

- Global analysis, giving the values of stress at the joints under these environmental load combinations

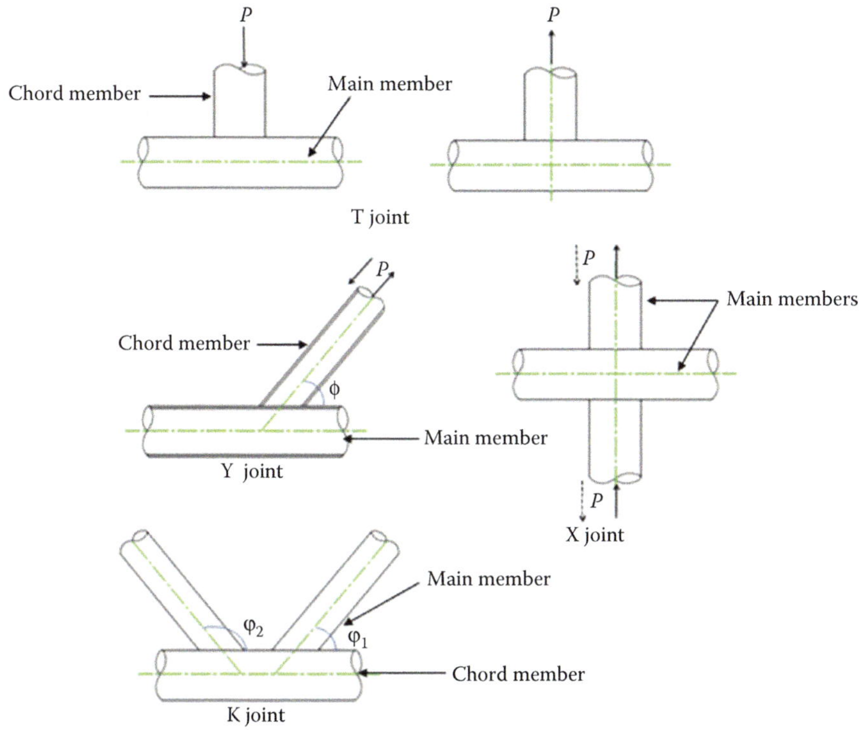

FIGURE 2.22 Types of tubular joints.

- Detailed local shell analysis of the joint under these nominal stresses, which lead to hot spot stresses
- Detailed 3D analysis near the weld of the tubular joints, because stresses are several times higher near the welds at tubular joints; they can be even higher than the yield strength

Local stress calculations are necessary for fatigue analysis of the joint. They do not provide a practical measure of the ultimate strength of the joint under static loads because tubular connections have higher reserve strength even after yield point is reached. Therefore, designers use design rules, which are based on empirical data derived from series of experimental investigations. Figure 2.22 shows typical tubular joints that are common in marine structural systems.

2.13.1 Ultimate Strength of T Joints in Compression

Figure 2.23 shows the T joint in compression. The T joint under compression is more critical than the T joint in tension because it changes the curvature of the chord significantly. Further, failure of the T joint in compression is a combination of plastic yield with stability problems in the chord. In tension, the chord may fail either by

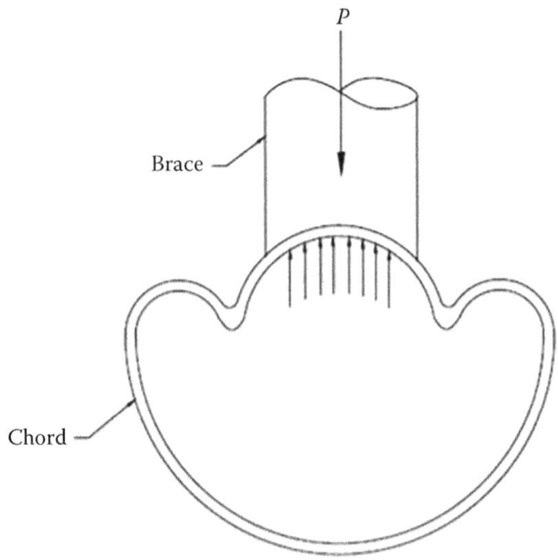

FIGURE 2.23 T joint under compression.

plastic yield due to tension or by rupture caused by the shear stresses exceeding the material strength.

A typical ring model, which is used to estimate the ultimate load-carrying capacity of a T joint in compression, is shown in Figure 2.24.

With reference to Figure 2.24, the chord ring is supported by the shear stress with q intensity/unit length. Ultimate load capacity of the T joint in compression is given by

$$P_{UC} = 7.5\beta\sqrt{\gamma}T^2\sigma_y \tag{2.45}$$

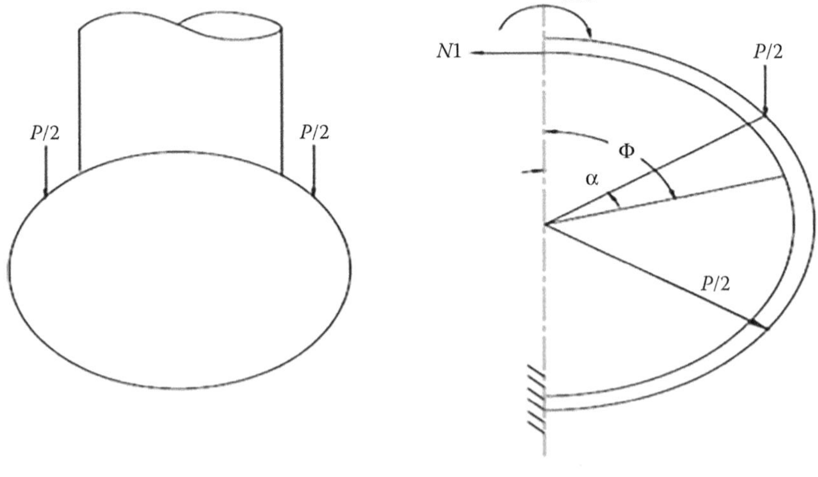

FIGURE 2.24 Ring model for T joint under compression.

where r is the brace radius, R is the chord radius, t is the brace thickness, and T is the chord thickness. The following ratios are valid for the T joint:

$$\beta = \frac{r}{R}$$

$$\gamma = \frac{R}{T}$$

Equation 2.45 is valid for the following range of parameters:

$$0.25 \leq \beta \leq 0.85$$

$$10 \leq \gamma \leq 20$$

It is to be noted here that the above ratios are based on experimental observations.

2.13.2 Ultimate Strength of T Joints in Tension

The ultimate strength of a T joint in tension is given by

$$P_{ut}^{T} = (2.3 + 6\beta)\sqrt{\gamma}T^{2}\sigma_{y} \tag{2.46}$$

The validity of the range of parameters is as follows:

$$0.2 \leq \beta \leq 1.0$$

$$9 \leq \gamma \leq 3.0$$

Combining Equations 2.45 and 2.46, we get

$$\frac{P_{ut}^{T}}{P_{uc}^{T}} = \frac{(2.3 + 6\beta)}{7.5\beta} \tag{2.47}$$

2.13.3 Ultimate Strength of Y Joints in Compression

The ultimate capacity of Y joints is generally expressed in terms of punching shear stress. Punching shear stress is the mean shear stress in the intersection line between the brace and the chord. Punching shear stress is a hypothetical parameter because Y joints actually do not fail by shear but fail by the formation of a plastic hinge at the joint under compressive loads. A typical Y joint is shown in Figure 2.25. Based on the figure, the vertical component of the compressive load is given by

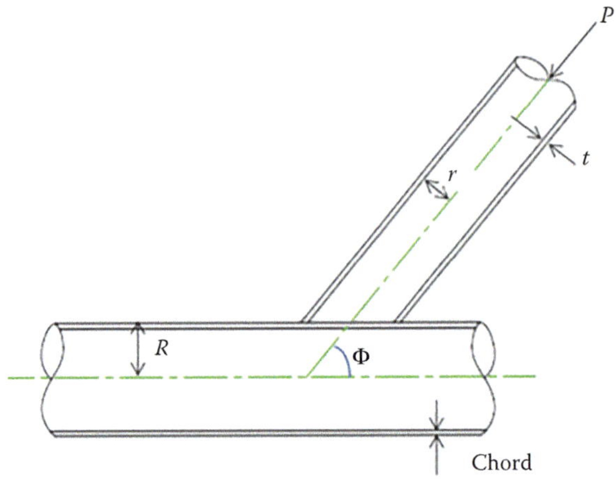

FIGURE 2.25 Y joint in compression.

$$P_V = P_{UC}^{Y} \sin \Phi$$

$$P_{UC}^{Y} = \frac{P_{UC}^{T}}{\sin \Phi} = \frac{7.5\beta\sqrt{r}T^2\sigma_y}{\sin \Phi} \tag{2.48}$$

The validity of the range of parameters is given by

$$0.25 \le \beta \le 0.85$$
$$10 \le \gamma \le 20$$
$$30° \le \Phi \le 90°$$

2.13.4 Ultimate Strength of Y Joints in Tension

The ultimate strength of a Y joint in tension is given by

$$P_{ut}^{Y} = \frac{P_{ut}^{T}}{\sin \Phi}$$

$$P_{ut}^{Y} = \frac{(2.3 + 6\beta)\sqrt{r}T^2\sigma_y}{\sin \Phi} \tag{2.49}$$

The validity of the range of parameters is given by

$$0.20 \le \beta \le 0.85$$
$$10 \le \gamma \le 20$$
$$30° \le \Phi \le 90°$$

Based on experimental investigations, it can be seen that the ultimate capacity of Y joints in tension, estimated from the above empirical equations, are higher. Hence, a correction factor, as suggested by DNV rules, is applied, as given below.

The correction factor is given by

$$c = \frac{1 + \sin \Phi}{2 \sin \Phi} \geq 1.0$$

$$\text{For } \Phi = 30°, C = 1.5$$

$$\text{For } \Phi = 90°, C = 1.0$$

The modified capacity of Y joints in tension is given by

$$P_{ut}^{y} = (2.5 + \sigma\beta)\sqrt{\gamma}T^2\sigma_y \left\{ \frac{1 + \sin \Phi}{2 \sin \Phi} \right\} \tag{2.50}$$

2.13.5 Ultimate Strength of K Joints

For a typical one-sided K joint, it is a common design practice that the sum of forces acting normal to the chord axis is set to *zero*. For a cross-type K joint, the component normal to the chord axis is transferred to the brace on the opposite side.

For a one-sided K joint, the ultimate strength is given by

$$P_{uc}^{K} = P_{uc}^{Y} \{ f_1(\varphi_1) f_2(\beta_1) f_3(g) \}$$

where φ_1 is the compression brace angle, φ_2 is the tension brace angle, a is the gap between the braces, f is eccentricity of the resultant force to chord axis, D is the chord diameter, g (= a/D) and e (= f/D) are design parameters, and f_1, f_2, f_3 are functions that take care of the influence of inclination, relative dimension of the compression brace, and chord gap, respectively. These functions are given by

$$\begin{aligned} f_1(\phi_1) &= 1 - 0.26 \cos^2 \phi_1 \\ f_2(\beta_1) &= \frac{1 + 6.1\beta_1}{4.2\beta_1} \\ f_3(g) &= \frac{2.4 + 1.8g}{2.4 + 7g} \end{aligned} \tag{2.51}$$

The range of validity of the parameters is given by

$$0.25 \leq \beta \leq 0.85$$

$$10 \leq \gamma \leq 55$$

$$0 \leq g \leq 0.6$$

$$30° \leq \Phi_{1,}\Phi_2 \leq 90°$$

2.13.6 Ultimate Strength of X Joints

The ultimate strength of X joints is compared to that of T joints to derive the expression of their strength. In X joints, the chord acts as a ring that transforms the forces from one brace to the other on the opposite side, whereas in T joints, load is transferred to the chord by shear stress. As a result, the capacity of X joints in compression is less than that of T joints in compression because ring stresses in the chord cause stability problems. The ultimate strength of X joint in compression is given by

$$P_{uc}^{X} = \frac{7.41}{(1.2 - \beta)} T^2 \sigma_y \tag{2.52}$$

The ultimate strength of X joints in tension is given by

$$P_{ut}^{X} = (1.48)(P_{uc}^{X}) \tag{2.53}$$

The validity of the parameter is given by

$$0.25 \leq \beta \leq 0.85$$

2.14 SHEAR CENTER

The shear center is the point of intersection of the longitudinal axis of the member and the line of action of transverse loads. In marine structures, members are usually thin and asymmetric. Generally, such cross sections are good in bending but weak in torsion. As shown in the figure, when the axis of the transverse load does not coincide with the centroid of the cross section, it induces an additional twisting moment about the shear center. If the cross section has both axes of symmetry, their intersection will be the shear center.

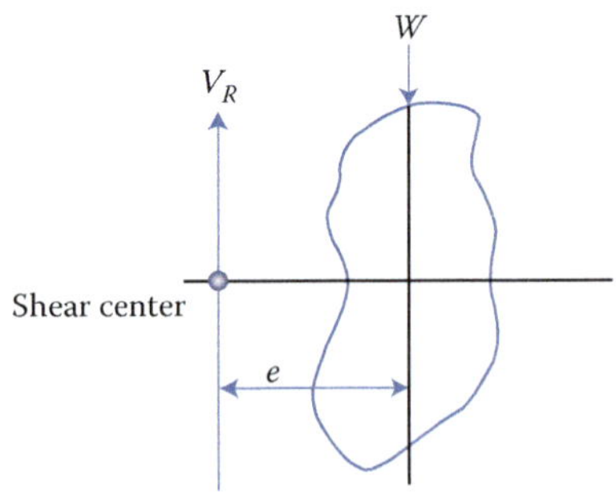

Let us illustrate the location of the shear center with an example. Consider the cross section shown below.

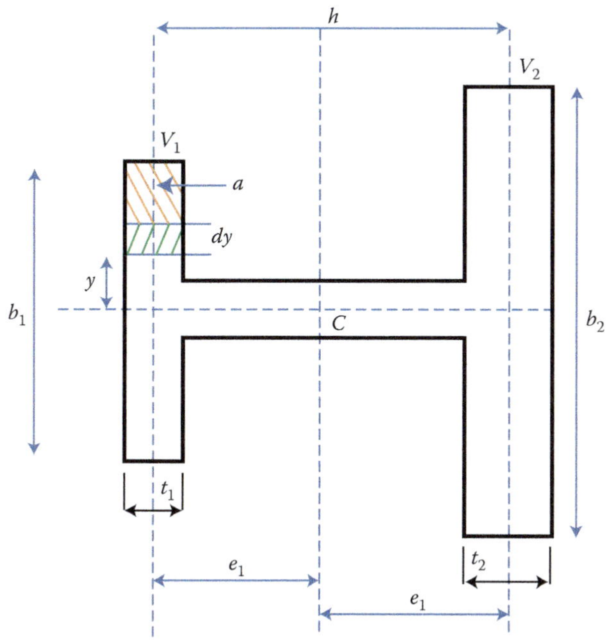

Total shear, $V = V_1 + V_2$

To find V_1, let us consider an element area, as shown in the figure. The intensity of shear stress is given by

$$\tau = \frac{Va\bar{y}}{Ib} = \frac{Wa\bar{y}}{It_1}$$

$$A = \left(\frac{b_1}{2} - y\right)t_1$$

$$\bar{y} = y + \frac{1}{2}\left(\frac{b_1}{2} - y\right) = \frac{b_1}{4} + \frac{y}{2}$$

$$\tau = \frac{W}{It_1}\left[\frac{b_1}{2} - y\right]t_1\left[\frac{y}{2} + \frac{b_1}{4}\right]$$

$$\tau = \frac{W}{2I}\left(\frac{b_1^2}{4} - y^2\right)$$

where I is the moment of inertia of the entire cross section, about the *X–X* axis.

The shear force in the left flange is given by

$$
\begin{aligned}
V_1 &= \int \tau \, da \\
&= \frac{W}{2I} \int_{-b_1/2}^{b_1/2} \left(\frac{b_1^2}{4} - y^2 \right) t_1 \, dy \\
&= \frac{W}{2I} \int_{-b_1/2}^{b_1/2} \left(\frac{b_1^2}{4} t_1 - y^2 t_1 \right) dy \\
&= \frac{W}{I} t_1 \left(\frac{b_1^3}{12} \right) \\
V_1 &= \frac{W}{I} I_1
\end{aligned}
$$

where

$$\left\{ I_1 = \frac{t_1 b_1^3}{12} \right\}$$

Similarly, the shear force in the right flange is given by

$$V_2 = \frac{W}{I} I_2$$

where

$$\left\{ I_2 = \frac{t_2 b_2^3}{12} \right\}$$

Now, the total shear (by neglecting that of the web) is given by

$$V = V_1 + V_2 = \frac{W}{I}(I_1 + I_2)$$

for $I = I_1 + I_2$, then the following relationships hold good:

$$V_1 e_1 = V_2 e_2$$

$$V_1 = \frac{W}{I}(I_1)$$

Since the moment of inertia of the section is known, then, by using the above relationship, the shear center of the section can be computed.

2.14.1 Alternate Method to Find the Shear Center

The shear center can also be located by an alternate method. Let us consider a channel section, shown in the figure.

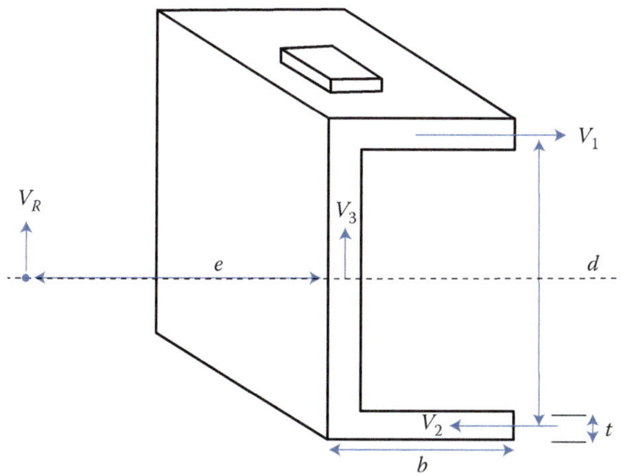

a. To find the shear stress in the flange:

$$H = \int \frac{M}{I} yda$$

$$H + dh = \int \frac{M + dM}{I} yda$$

The unbalanced longitudinal shear force is given by

$$dH = \int \frac{dM}{I} yda$$

For equilibrium, the element shear stress must oppose this unbalanced force. Let τ be the shear stress, then

$$\tau(tdz) = \int_x^b \frac{dM}{I} yda$$

$$\tau = \frac{dM}{dZ} \frac{1}{It} a\bar{y}$$

$$\tau = \frac{Va\bar{y}}{It}$$

Let V_1, V_2, V_3 be the shear resistances offered by the flanges and web, as indicated in the figure. V_R is the resultant shear, which will act through the shear center. In order to avoid twisting moment in the cross section, V must be coplanar and collinear with V_R.

$$V_1 = \int \tau\, da = \int \frac{Va\bar{y}}{It}\, da$$

$$a = tx$$

$$\bar{y} = \frac{d}{2}$$

$$da = (dx)t$$

Then

$$V_1 = \int_x^b \frac{V}{It}(xt)\frac{d}{2}\,dx(t)$$

$$= \int_x^b \frac{V}{It}(xt)^2\,\frac{d}{2}\,dx$$

$$= \frac{Vt^2 d}{2It}\int_x^b x\,dx$$

$$V_1 = \frac{V\,dt^2}{2It}\left(\frac{b^2}{2} - \frac{x^2}{2}\right)$$

Considering a section at $x = 0$, we get

$$V_1 = \frac{Vdt^2}{2It}\left(\frac{b^2}{2}\right)$$

Similarly, V_2 can be computed. By substituting, we get

$$V_R(e) = V_1\frac{d}{2} + V_2\frac{d}{2} \text{ (by ignoring } V_3)$$

The shear center can be computed by solving the following equation:

$$V_R(e) = V_1\frac{d}{2} + V_2\frac{d}{2}$$

2.14.2 Shear Center of the Special Section

Let us consider a section as shown in the figure.

By symmetry, $V_1 = V_5$ and $V_2 = V_4$.

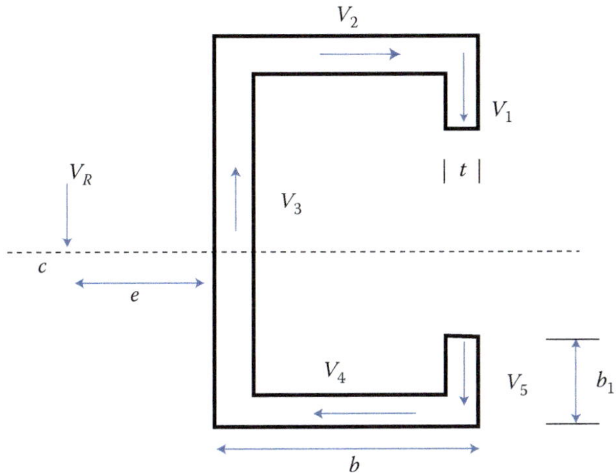

Shear force V_1 is given by

$$V_1 = \int_0^{b_1} \frac{Va\bar{y}}{It} da$$

$$a\bar{y} = Q = (tz)\left(\frac{h}{2} - b_1 + \frac{z}{2}\right)$$

$$V_1 = \int_0^{b_1} \frac{V}{It}(tz)\left(\frac{h}{2} - b_1 + \frac{z}{2}\right) dz(t) = \frac{Vtb_1^2}{I}\left\{\frac{h}{4} - \frac{b_1}{3}\right\}$$

Shear force V_2 is given by

$$V_2 = \int \frac{Va\bar{y}}{It} da$$

$$da = tdx$$

$$a\bar{y} = Q$$

$$= (xt)\frac{h}{2}$$

$$V_2 = \int_0^{b} \frac{V}{It}(xt)\frac{h}{2} t\, dx$$

$$V_2 = \frac{Vt}{I}\frac{b^2h}{4}$$

We know that $V_1 = V_5$, $V_2 = V_4$.

Taking the moment about the shear center, we get

$$Ve = (V_1 + V_5)b + (V_2 + V_4)\frac{h}{2}$$

By solving the above equation, the shear center can be determined.

2.15 PLASTIC MOMENT-CARRYING CAPACITY OF SECTIONS UNDER COMBINATIONS OF LOADS

Generally, structural sections used for marine structures should be checked for the buckling effect. Sections are classified as compact and noncompact sections to differentiate whether buckling failure will preclude the failure under the combination of loads. For compact sections, local buckling can be disregarded because yielding will preclude buckling. On the other hand, for noncompact sections, local buckling should be considered in the design. The following are the set of equations that can be used to check the compact limit of various cross sections that are commonly used in marine structures.

2.15.1 Tubular Elements

A typical tubular element is shown in the figure. A tubular element is said to be compact if the following condition is satisfied:

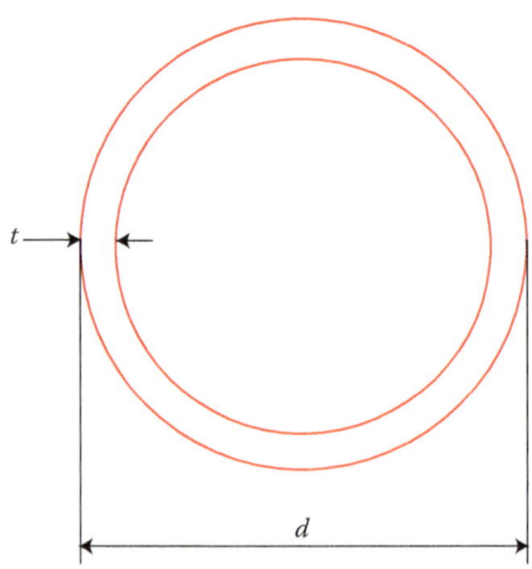

$$\frac{d}{t} \leq \frac{E}{9\sigma_{yp}}$$

2.15.2 Box Section

A typical box section is shown. The section is considered compact if the following condition is satisfied:

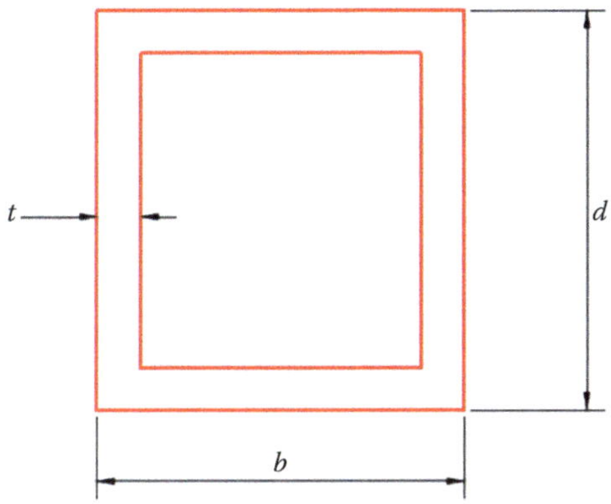

$$\frac{b}{t}, \frac{d}{t} \leq 1.5\sqrt{\frac{E}{\sigma_{yp}}}$$

2.15.3 Welded Box-Type Section

A typical welded box-type section is shown. The section is considered compact if both the following equations are satisfied:

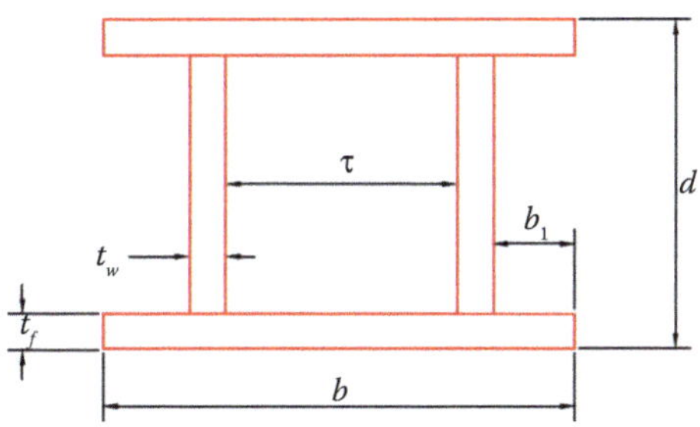

$$\frac{a}{t_f}, \frac{d}{t_w} \leq 1.5\sqrt{\frac{E}{\sigma_{yp}}}$$

$$\frac{b_1}{t_f} \leq 0.4\frac{E}{\sqrt{\sigma_{yp}}}$$

2.15.4 I Section

A typical I section is shown. The section is considered compact if both the following equations are satisfied:

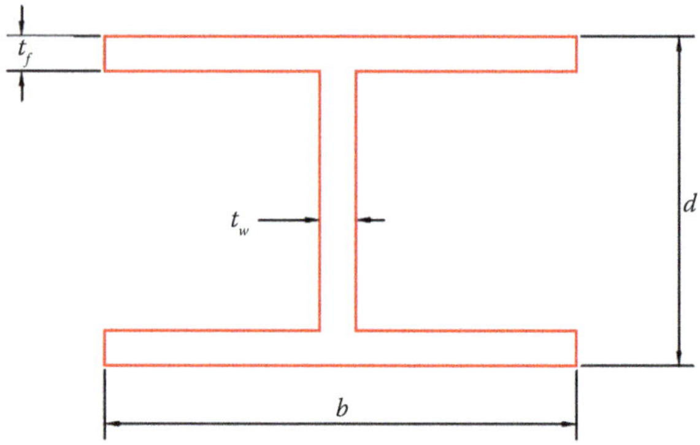

$$\frac{d}{t_w} \leq 1.5\sqrt{\frac{E}{\sigma_{yp}}}$$

$$\frac{b}{t_f} \leq 0.8\sqrt{\frac{E}{\sigma_{yp}}}$$

2.15.5 Channel Section

A typical channel section is shown. The section is considered compact if both the following equations are satisfied:

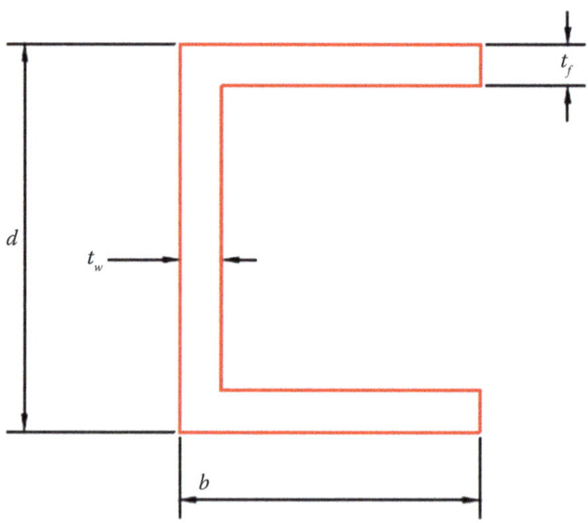

$$\frac{d}{t_w} \leq 1.5\sqrt{\frac{E}{\sigma_{yp}}}$$

$$\frac{b}{t_f} \le 0.4\sqrt{\frac{E}{\sigma_{yp}}}$$

2.15.6 Inverted T Section

A typical inverted T section is shown. The section is considered compact if both the following equations are satisfied:

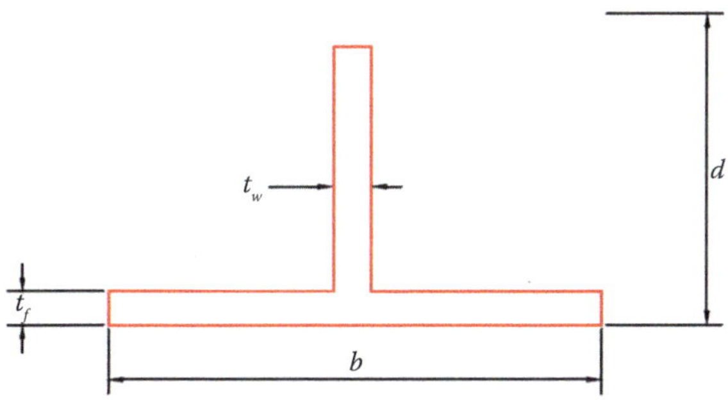

$$\frac{d}{t_w} \le 0.4\sqrt{\frac{E}{\sigma_{yp}}}$$

$$\frac{b}{t_f} \le 0.8\sqrt{\frac{E}{\sigma_{yp}}}$$

2.15.7 Double Angle Section, Back to Back

A typical double angle section, placed back to back, is shown. The section is considered compact if both the following equations are satisfied:

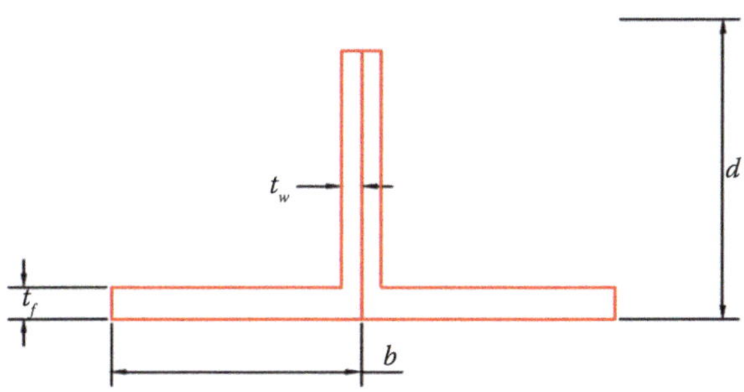

$$\frac{d}{t_w} \le 0.4\sqrt{\frac{E}{\sigma_{yp}}}$$

$$\frac{b}{t_f} \leq 0.4\sqrt{\frac{E}{\sigma_{yp}}}$$

2.16 PLASTIC CAPACITY OF SECTIONS UNDER AXIAL LOADS

For sections under axial tension and compression, following relationship holds good:

$$N_p = A\sigma_{yp}$$

where A is the effective area of cross section. The shear capacity of the section is computed based on the orientation of the member with respect to the direction of shear force. A typical I section, with two possible orientations, is shown. The shear capacity, for each case, is given by

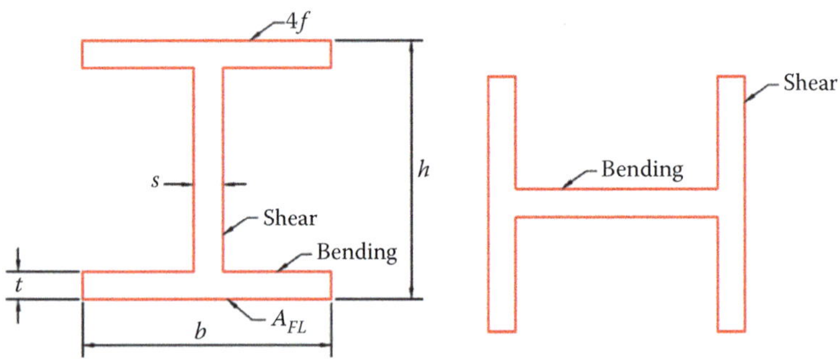

(i) orientation (a)
(ii) orientation (b)

$V_p = A_w \frac{\sigma_{yp}}{\sqrt{3}}$ (for orientation a, where minor axis is in line with the direction of load)

$V_p = (A_{f1} + A_{f2})\frac{\sigma_{yp}}{\sqrt{3}}$ (for orientation b, where major axis is in line with the direction of load)

where

$$A_w = (h - 2t)s$$

$$A_{f1} = A_{f2} = (bt)$$

2.17 TORSION CAPACITY

The torsion capacity of sections used in marine structures depends on the whether the section is fully plasticized or not. Two common types of section used in the design are (i) open thin-walled sections and (ii) closed thin-walled sections.

2.17.1 Open Sections

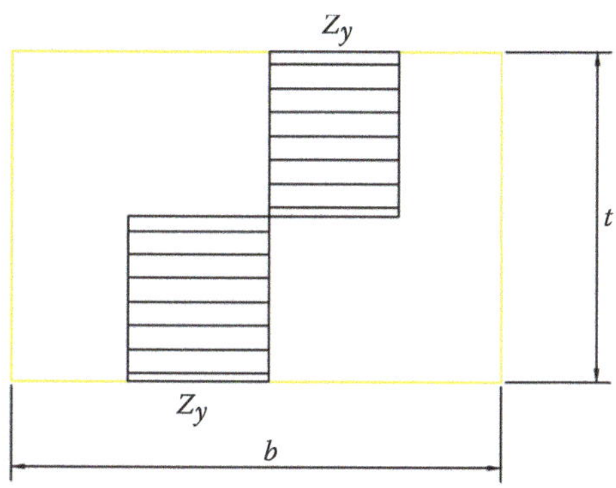

When the section becomes fully plasticized, the shear stress will reach the yield strength value. The shear force in each direction is given by

$$s = b * \frac{t}{2} * \tau_y$$

The torsional moment that is required to balance this shear is given by

$$T_p = s * \frac{t}{2}$$

$$T_p = \left(\frac{bt}{2}\tau_y\right)\frac{t}{2}$$

$$T_p = \frac{bt^2}{4}\tau_y$$

According to von Mises yielding criterion, torsional capacity is given by

$$\tau_y = \frac{\sigma_{yp}}{\sqrt{3}}$$

$$T_p = \frac{bt^2}{4}\frac{\sigma_{yp}}{\sqrt{3}}$$

2.17.2 I Sections

The torsion capacity of I sections is computed by summing up the capacities of the individual components (web and flanges), as given below:

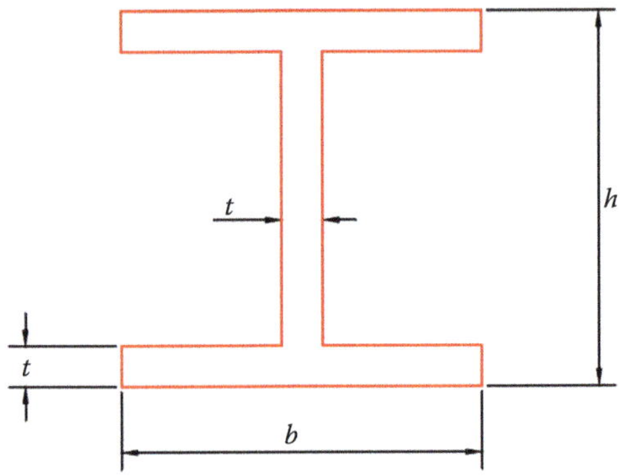

$$T_p = \sum_i \frac{b_i t_i^2}{4} \frac{\sigma_{yp}}{\sqrt{3}}$$

$$T_p = \frac{\sigma_{yp}}{\sqrt{3}} * \frac{1}{4}\left\{(bt^2(2) + (h - 2t)t^2)\right\}$$

2.17.3 Closed Sections

A typical closed section is shown in the figure. In the plastic range, shear stress takes the value of yield strength throughout the wall thickness. The torsion capacity of the section is given by

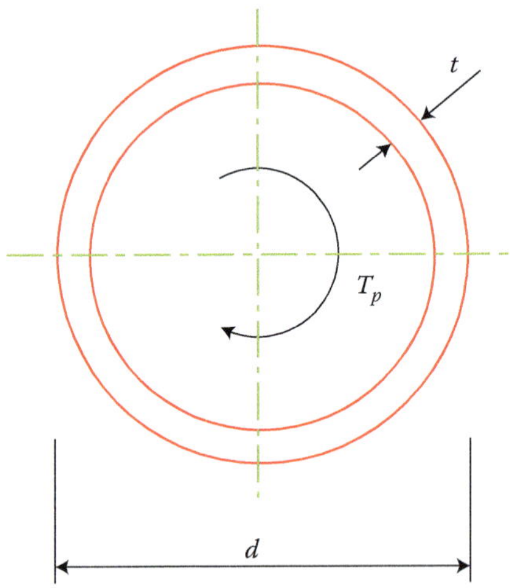

$$T_p = (\pi dt)\tau_y \left(\frac{d}{2}\right) = \frac{\pi d^2}{2} t\tau_y$$

Applying von Mises criteria of failure, we get

$$T_p = \frac{\pi d^2}{2} t \frac{\sigma_{yp}}{\sqrt{3}}$$

$$= 2At\frac{\sigma_{yp}}{\sqrt{3}}$$

where A is the area enclosed by the ring.

2.18 PLASTIC CAPACITY OF SECTIONS UNDER BENDING AND AXIAL LOAD

2.18.1 Rectangular Cross Section

A typical rectangular cross section subjected to bending and axial load is shown in the figure.

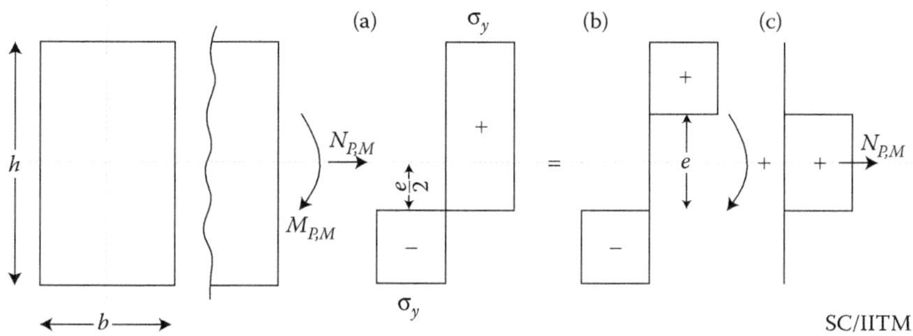

Let $M_{P,N}$ denote the reduced plastic moment-carrying capacity due to the presence of axial load N, and $N_{P,M}$ denote the reduced axial load capacity of the section due to the P–M interaction. To estimate the plastic capacity, the strength of the section is divided into two: (i) the pure bending case and (ii) the pure axial load case, as shown in the figure.

$$M_{P,N} = \frac{bh^2}{4}\sigma_y - \frac{be^2}{4}\sigma_y$$

$$= M_P\left[1 - \frac{e^2}{h^2}\right]$$

$$N_{P,M} = be\sigma_y = bh\sigma_y\frac{e}{h}$$

$$N_{P,M} = N_P\frac{e}{h}$$

Combining both the above equations for $M_{P,N}$ and $N_{P,M}$, we get

$$\frac{M_{P,N}}{M_P} + \left[\frac{N_{P,M}}{N_P}\right]^2 = 1$$

The above equation will result in an interaction diagram for rectangular cross sections, as shown in the figure.

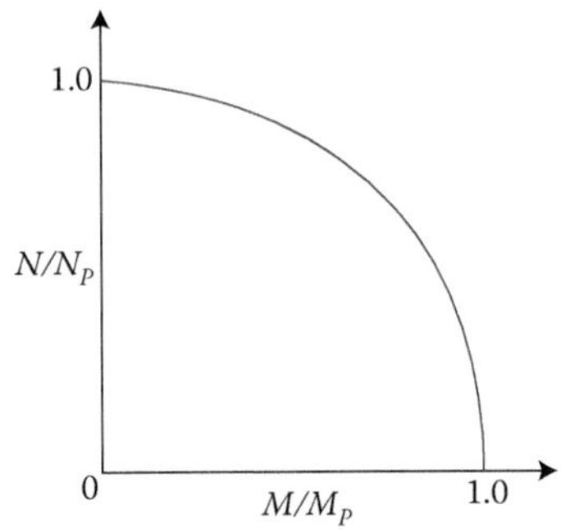

SC/IITM

2.18.2 I Sections

The stress distribution diagram for I section is shown.

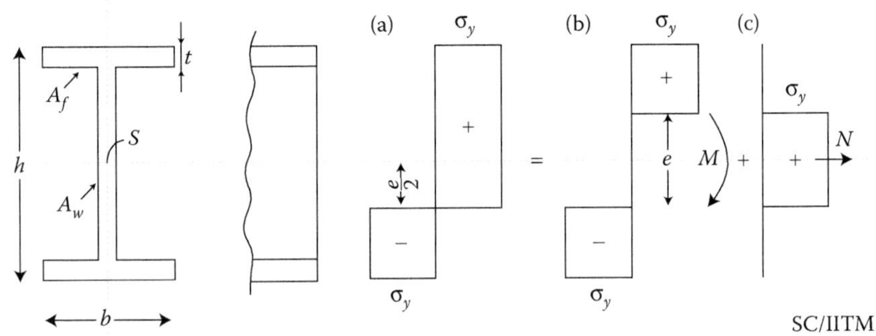

SC/IITM

Let us assume that the axis of *zero stress* lies at a distance of (0.5e) from the centroidal axis of the section, as shown in the figure. The bending moment is given by

$$M = M_P - \frac{se^2}{4}\sigma_y$$

The axial force is given by

$$N = es\sigma_y$$

Combining the above equations, we get

$$M = M_P - \frac{N^2}{N_P^2}\left\{\frac{A}{A_w}\right\}^2 \frac{sh_w^2}{4}\sigma_{yp}$$

$$1 = \frac{M}{M_P} + \frac{N^2}{N_P^2}\left\{\frac{A}{A_w}\right\}^2 \frac{z_w}{z_p}$$

where z_w is the plastic section modulus of the web and z_p is the plastic section modulus of the complete section. For I sections, plastic section modulus of the complete section is given by

$$\begin{aligned} z_p &\approx A_f(h-t) + z_{\text{web}} \\ &\approx \frac{A - A_w}{2}h_w + \frac{h_w A_w}{4} \\ &= z_w\left\{2\left(\frac{A}{A_w}\right) - 1\right\} \end{aligned}$$

The interaction equation under the combined action of bending moment and axial forces is given by

$$\frac{M}{M_P} + \left[\frac{N}{N_P}\right]^2 \left\{\frac{1}{2\left(\frac{A_w}{A}\right) - \left(\frac{A_w}{A}\right)^2}\right\} = 1$$

The above equation is valid if axial load N is less than the web capacity ($A_w\ \sigma_{yp}$), or the following condition applies:

$$\left(\frac{N}{N_P}\right) \le \left(\frac{A_w}{A}\right)$$

For axial force N greater than the web capacity, the following is valid:

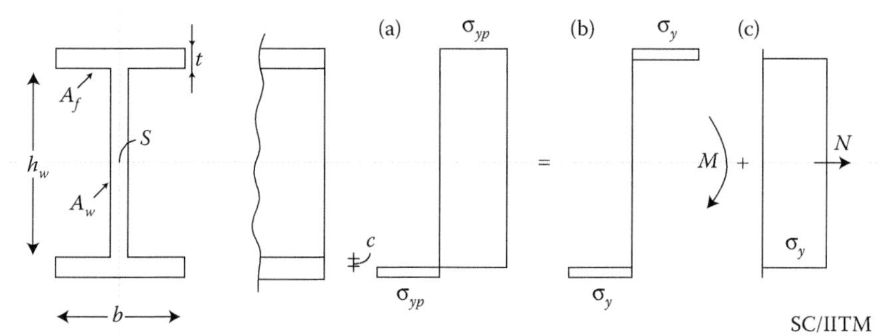

With reference to the figure, let c denote the distance of the inner flange from the axis of *zero stress*; then the following relationship holds good:

$$M = M_P - [bc(h_w + c) + z_w]\sigma_{yp}$$
$$N = (A_w + 2bc)\sigma_{yp}$$

For a thin-walled section, $c \ll h_w$, and hence by combining the above two equations, we get

$$M = M_P - \frac{Nh_w}{c} + z_w\sigma_{yp}$$

Also, $Z_P \cong z_w\left[2\frac{A}{A_w} - 1\right]$

$N_P = A \cdot \sigma_{yp}$

The interaction form is given by

$$\frac{M}{M_P} + \frac{N}{N_P}\left\{\frac{1}{(1-(\mathrm{A_w}/2\mathrm{A}))} - \frac{\mathrm{A_w}/\mathrm{A}}{2\,(1-(\mathrm{A_w}/2\mathrm{A}))}\right\} = 1$$

where the following condition applies:

$$\left(\frac{N}{N_P}\right) \geq \left(\frac{A_w}{A}\right)$$

The equation of the interaction diagram can also be rewritten in a modified form, as given below:

$$\frac{M}{M_P}\left[1 - \frac{A_w}{2A}\right] + \frac{N}{N_P} = 1$$

A plot of the interaction diagram is shown.

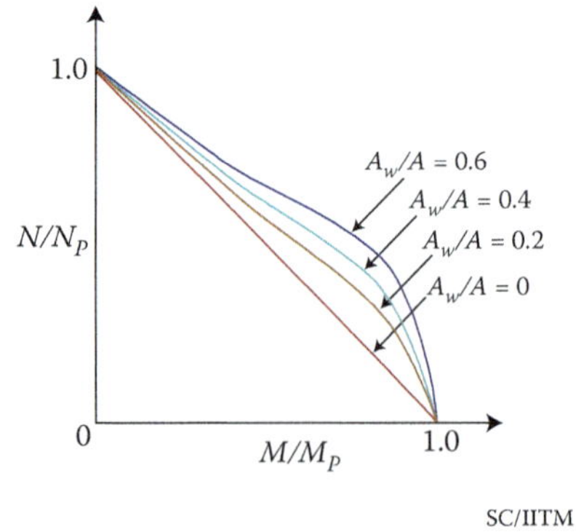

2.19 STRUCTURAL DESIGN OF MEMBERS USING AMERICAN BUREAU OF SHIPPING CODE (2008)

(From ABS. 2008. *Guide for Buckling and Ultimate Strength Assessment of Offshore Structures*. American Bureau of Shipping, New York.)

2.19.1 Example Problem: Offshore Tubular Members

Examine the safety of an offshore tubular member using ABS code. The following data are available:

$D = 600$ mm
Thickness, $t = 12$ mm
$E = 2 \times 10^5$ N/mm^2
Yield stress of the material, $\sigma_0 = 240$ N/mm^2
Length of the member, $l = 4$ m
Effective length factor, $K = 0.8$
Axial compression, $Pc = 1$ MN
The platform is subjected to severe storm conditions.

Solution
Step # 1: To determine basic design parameters:

a. Cross-sectional area, A

$$A = \frac{\pi}{4}\left(D_o^2 - D_i^2\right)$$

$$\frac{\pi}{4}(600^2 - 576^2) = 22167.08 \text{ mm}^2$$

Axial compressive stress is given by

$$\frac{P}{A} = \frac{1 \times 10^6}{22167.08} = 45.12 \text{ N/mm}^2$$

b. Moment of inertia, I

$$I = \frac{\pi}{64}\left(D_o^4 - D_i^4\right)$$

$$\frac{\pi}{64}(600^4 - 576^4) = 0.96 \times 10^9 \text{ mm}^4$$

c. Radius of gyration, r_{yy}

$$r_{yy} = \sqrt{\frac{I}{A}} = \sqrt{\frac{0.96 \times 10^9}{22167.08}} = 208.10 \text{ mm}$$

d. Effective length, $Kl = (0.8 \times 4000) = 3200$ mm

e. Polar moment of inertia, I_0

$$I_0 = \frac{\pi}{32}\left(D_o^4 - D_i^4\right)$$

$$\frac{\pi}{32}(600^4 - 576^4) = 1.92 \times 10^9 \text{ mm}^4$$

f. St. Venant constant for tubular member, k

$$k = \frac{\pi}{4}(D - t)^3 t$$

$$\frac{\pi}{4}(600 - 12)^3 \times 12 = 1.916 \times 10^9 \text{ mm}^4$$

g. Euler's buckling stress, $\sigma_E\eta$

$$\frac{\pi^2 E}{[k\ell / r_{yy}]^2} = \frac{\pi^2 \times 2 \times 10^5}{[3200/208.10]^2} = 8347.84\,\text{N/mm}^2$$

h. Warping constant, $\Gamma = 0$ for tubular members
i. d_{cs} is the distance of centroid and shear center along the major axis, $d_{cs} = 0$ for the present case

Step # 2: To determine compact limit for tubular members:
It is necessary to establish the compact limit for tubular members because local buckling similar to that of plate buckling, shell buckling, etc. need not be considered for compact sections; in such cases, yielding precedes buckling. For noncompact sections, local buckling should be taken into account. Tubular sections are said to be compact if the following equation is satisfied:

$$\frac{D}{t} \le \frac{E}{9\sigma_0}$$

$$\frac{600}{50} = 12$$

$$\frac{E}{9\sigma_0} = \frac{2 \times 10^5}{9 \times 240} = 92.59$$

Since the above ratio is satisfied, the tubular section is compact; yielding precedes buckling.

Step # 3: To determine the adjustment factor for computing maximum allowable strength utilization factor, η:

The maximum allowable strength utilization factor depends on the loading conditions, namely: (a) for static loading, the utilization factor is (0.6 ψ) and (b) for a severe storm case, the factor is (0.8 ψ), where ψ is the adjustment factor. We shall now evaluate the adjustment factor, ψ. For axial compression that will result in column buckling or torsion buckling, the adjustment factor is given by the following expression:

$$\psi = 0.87(\text{if } \sigma_{EA} \leq P_r\sigma_0)$$
$$= 1 - 0.13\sqrt{\frac{P_r\sigma_0}{\sigma_{EA}}}(\text{if } \sigma_{EA} > P_r\sigma_0)$$

where σ_{EA} is the elastic buckling stress about the minor axis, P_r is the proportional elastic limit, and σ_0 is the yield stress. The proportional elastic limit is a material characteristic and can be obtained from the stress–strain curve of the material. In the case of absence of data, this value can be taken as 0.6. The above equation demands computation of elastic buckling stress about the minor axis.

Step # 4: To compute elastic buckling stress, σ_{EA}:
The elastic buckling stress is given by the least of the solutions to the following quadratic equation:

$$\frac{I_0}{A}[\sigma_{EA} - \sigma_{E\eta}][\sigma_{EA} - \sigma_{ET}] - \left[\sigma_{EA}^2 - d_{cs}^2\right] = 0$$

where σ_{ET} is the elastic torsion buckling stress and is given by

$$\sigma_{ET} = \frac{EK}{2.6I_0} + \left[\frac{\pi}{k\ell}\right]^2 \frac{E\Gamma}{I_0}$$

$$\sigma_{ET} = \frac{2\times10^5\times1.916\times10^9}{2.6\times1.92\times10^9} + \left[\frac{\pi}{3200}\right]^2 \frac{2\times10^5\times\Gamma}{1.92\times10^9}$$
$$= 0.769\times10^5 \text{ N/mm}^2$$

Substituting, we get

$$\frac{I_0}{A}[\sigma_{EA} - \sigma_{E\eta}][\sigma_{EA} - \sigma_{ET}] - [\sigma_{EA}^2 - d_{cs}^2] = 0$$

$$\frac{1.92\times10^9}{22167.08}[\sigma_{EA} - 8347.84][\sigma_{EA} - 0.769\times10^5] = 0$$

Solving the above equation for the unknown and choosing the lowest root of the above equation, we get

$$\sigma_{EA} = 8348.59 \text{ N/mm}^2$$

Step # 5: To compute the adjustment factor:
Checking for the condition, $\sigma_{EA} > P_r\,\sigma_0$, we get

$$8348.59 > (0.6 \times 240\ \text{N/mm}^2)$$

Hence, the adjustment factor for the present case will be given by

$$\psi = 1 - 0.13\sqrt{\frac{P_r\sigma_0}{\sigma_{EA}}}\,(\text{if } \sigma_{EA} > P_r\sigma_0)$$

$$= 1 - 0.13\sqrt{\frac{0.6 \times 240}{8348.59}} = 0.131$$

The strength utilization factor (for the severe storm case, as in the present case), η, is given by

$$\eta = (0.8\,\psi) = 0.8 \times 0.131 = 0.1048$$

Step # 6: To compute the critical buckling stress, σ_{CA}:

$$\sigma_{CA} = \begin{cases} \sigma_{EA} \text{ if } \sigma_{EA} \le P_r\sigma_F \\ \sigma_F\left[1 - P_r(1 - P_r)\dfrac{\sigma_F}{\sigma_{EA}}\right] \text{if } \sigma_{EA} > P_r\sigma_F \end{cases}$$

σ_F is the minimum yield point stress for compact sections. In case of absence of such data, it can be taken as equal to the yield stress specified.

Hence, $\sigma_F = \sigma_0 = 240$ N/mm^2

Now, $\sigma_{EA} > P_r\,\sigma_F$ and hence

$$\sigma_{CA} = \sigma_F\left[1 - P_r(1 - P_r)\frac{\sigma_F}{\sigma_{EA}}\right]$$

$$= 240\left[1 - 0.6(1 - 0.6)\frac{240}{8348.59}\right] = 238.32\ \text{N/mm}^2$$

Step # 7: Check for axial compression:
In the case of tubular members subjected to axial compression where the member is compact, the failure may occur due to flexural or torsion buckling. The buckling limit is given by the following expression:

$$\left[\frac{\sigma_A}{\eta\sigma_{CA}}\right] \le 1 \quad \text{for safe}$$

$$\left[\frac{45.12}{0.1048 \times 238.32}\right] = 1.807 > 1; \quad \text{unsafe}$$

The tubular member is assessed as unsafe in axial compression; it may fail in column buckling.

2.19.2 Example Problem: Tubular Members under Combination of Forces

Examine the safety of the tubular member of an offshore platform for normal operations. Design data useful for the assessment are given below. Use ABS code.

$D = 1000$ mm
Thickness, $t = 15$ mm
$E = 2 \times 10^5$ N/mm^2
Yield stress of the material, $\sigma_0 = 240$ N/mm^2
Length of the member, $_i = 15$ m
Effective length factor, $K = 0.8$
Axial compression, $Pc = 1$ MN
Bending moment = 1000 kNm
The platform is subjected to normal operating conditions.

Solution
Step # 1: To determine basic design parameters:

a. Cross-sectional area, A

$$A = \frac{\pi}{4}(D_o^2 - D_i^2)$$

$$\frac{\pi}{4}(1000^2 - 970^2) = 46417.03 \text{ mm}^2$$

Axial compressive stress is given by

$$\frac{P}{A} = \frac{1 \times 10^6}{46417.03} = 21.54 \text{ N/mm}^2$$

b. Moment of *inertia*, I

$$I = \frac{\pi}{64}(D_o^4 - D_i^4)$$

$$\frac{\pi}{64}(1000^4 - 970^4) = 5.63 \times 10^9 \text{ mm}^4$$

c. Radius of gyration, r_{yy}

$$r_{yy} = \sqrt{\frac{I}{A}} = \sqrt{\frac{5.63 \times 10^9}{46417.03}} = 348.27 \text{ mm}$$

d. Effective length, $Kl = (0.8 \times 15{,}000) = 12{,}000$ mm
e. Polar moment of inertia, I_0

$$I_0 = \frac{\pi}{32}\left(D_o^4 - D_i^4\right)$$

$$\frac{\pi}{32}(1000^4 - 970^4) = 1.126 \times 10^{10} \text{ mm}^4$$

f. St. Venant constant for tubular member, k

$$k = \frac{\pi}{4}(D - t)^3 t$$

$$\frac{\pi}{4}(1000 - 15)^3 \times 15 = 1.126 \times 10^{10}\ \text{mm}^4$$

g. Euler's buckling stress, $\sigma_E \eta$

$$\frac{\pi^2 E}{\left[k\ell / r_{yy}\right]^2} = \frac{\pi^2 \times 2 \times 10^5}{\left[12{,}000/348.47\right]^2} = 1662.64\ \text{N/mm}^2$$

h. Warping constant $\Gamma = 0$ for tubular members
i. d_{cs} is the distance of the centroid and shear center along the major axis, $d_{cs} = 0$ for the present case

Step # 2: To determine compact limit for tubular members:
It is necessary to establish the compact limit for tubular members because local buckling similar to that of plate buckling, shell buckling, etc., need not be considered for compact sections; in such cases, yielding precedes buckling. For noncompact sections, local buckling should be taken into account. Tubular sections are said to be compact if the following equation is satisfied:

$$\frac{D}{t} \leq \frac{E}{9\sigma_0}$$

$$\frac{1000}{15} = 66.67$$

$$\frac{E}{9\sigma_0} = \frac{2 \times 10^5}{9 \times 240} = 92.59$$

Since the above ratio is satisfied, tubular section is compact; yielding precedes buckling.

Step # 3: To determine the adjustment factor for computing maximum allowable strength utilization factor, η:
Maximum allowable strength utilization factor depends on the loading conditions, namely: (a) for loading static loading, the utilization factor is (0.6 ψ) and (b) for the severe storm case, the factor is (0.8 ψ), where ψ is the adjustment factor. We shall now evaluate the adjustment factor, ψ. For axial compression that will result in column buckling or torsion buckling, the adjustment factor is given by the following expression:

$$\psi = 0.87(\text{if } \sigma_{EA} \le P_r\sigma_0)$$
$$= 1 - 0.13\sqrt{\frac{P_r\sigma_0}{\sigma_{EA}}}(\text{if } \sigma_{EA} > P_r\sigma_0)$$

where σ_{EA} is the elastic buckling stress about the minor axis, P_r is proportional elastic limit, and σ_0 is yield stress. The proportional elastic limit is a material characteristic and can be obtained from the stress–strain curve of the material. In case of absence of this data, the value can be taken as 0.6. The above equation demands computation of elastic buckling stress about the minor axis.

Step # 4: To compute elastic buckling stress, σ_{EA}:
The elastic buckling stress is given by the least of the solutions to the following quadratic equation:

$$\frac{I_0}{A}[\sigma_{EA} - \sigma_{E\eta}][\sigma_{EA} - \sigma_{ET}] - \left[\sigma_{EA}^2 - d_{cs}^2\right] = 0$$

where σ_{ET} is the elastic torsion buckling stress and is given by

$$\sigma_{ET} = \frac{EK}{2.6I_0} + \left[\frac{\pi}{k\ell}\right]^2 \frac{E\Gamma}{I_0}$$

$$\sigma_{ET} = \frac{2\times10^5 \times 1.126\times10^{10}}{2.6\times1.126\times10^{10}} + \left[\frac{\pi}{3200}\right]^2 \frac{2\times10^5\times\Gamma}{1.92\times10^9}$$
$$= 76{,}913.84 \text{ N/mm}^2$$

Substituting, we get

$$\frac{I_0}{A}[\sigma_{EA} - \sigma_{E\eta}][\sigma_{EA} - \sigma_{ET}] - \left[\sigma_{EA}^2 - d_{cs}^2\right] = 0$$

$$\frac{1.126\times10^{10}}{46417.03}[\sigma_{EA} - 1662.64][\sigma_{EA} - 76913.84] = 0$$

Solving the above equation for the unknown and choosing the lowest root of the above equation, we get

$$\sigma_{EA} = 1650.95 \text{ N/mm}^2$$

Step # 5: To compute the critical buckling stress, σ_{CA}:

$$\sigma_{CA} = \begin{cases} \sigma_{EA} & \text{if } \sigma_{EA} \le P_r\sigma_F \\ \sigma_F\left[1 - P_r(1-P_r)\dfrac{\sigma_F}{\sigma_{EA}}\right] & \text{if } \sigma_{EA} > P_r\sigma_F \end{cases}$$

σ_F is the minimum yield point stress for compact sections. In case of absence of such data, it can be taken as equal to the yield stress specified.

Hence,

$$\sigma_F = \sigma_0 = 240 \text{ N/mm}^2$$

Now

$$\sigma_{EA} > P_r\,\sigma_F;\ (1650.95 > 0.6 \times 240)$$

Hence, we get

$$\sigma_{CA} = \sigma_F\left[1 - P_r(1 - P_r)\frac{\sigma_F}{\sigma_{EA}}\right]$$

$$= 240\left[1 - 0.6(1 - 0.6)\frac{240}{1650.95}\right] = 231.63 \text{ N/mm}^2$$

Step # 6: To check for the combined action of axial compression and bending moment: Tubular members subjected to axial compression and bending moment must satisfy the following equation:

$\dfrac{\sigma_A}{\sigma_{CA}} > 0.15$, then

$$\frac{\sigma_A}{\eta_1\sigma_{CA}} + \frac{1}{\eta_2}\left\{\left[\frac{1}{\sigma_{CBy}}\left(\frac{C_{My}\sigma_{by}}{1 - (\sigma_A/\eta_1\sigma_{Ey})}\right)\right]^2 + \left[\frac{1}{\sigma_{CBz}}\left(\frac{C_{Mz}\sigma_{bz}}{1 - (\sigma_A/\eta_1\sigma_{Ez})}\right)\right]^2\right\}^{0.5} \leq 1.0$$

$\dfrac{\sigma_A}{\sigma_{CA}} \leq 0.15$, then

$$\frac{\sigma_A}{\eta_1\sigma_{CA}} + \frac{1}{\eta_2}\left\{\left[\left(\frac{\sigma_{by}}{\sigma_{Cby}}\right)\right]^2 + \left[\left(\frac{\sigma_{bz}}{\sigma_{Cbz}}\right)\right]^2\right\}^{0.5} \leq 1.0$$

Now, for the present case,

$$\frac{\sigma_A}{\sigma_{CA}} = \frac{21,54}{231.63} = 0.093 < 0.15$$

Hence, the governing equation will be given by

$$\frac{\sigma_A}{\eta_1\sigma_{CA}} + \frac{1}{\eta_2}\left\{\left[\left(\frac{\sigma_{by}}{\sigma_{Cby}}\right)\right]^2 + \left[\left(\frac{\sigma_{bz}}{\sigma_{Cbz}}\right)\right]^2\right\}^{0.5} \leq 1.0$$

For axial compression, column buckling will be predominant in the present case, as the section is compact. Hence,

$$\eta_1 = 0.6\psi$$

$$\psi = 1 - 0.13\sqrt{\frac{P_r\sigma_0}{\sigma_{EA}}} \,(\text{for } \sigma_{EA} > P_r\sigma_0)$$

$$= 1 - 0.13\sqrt{\frac{0.6 \times 240}{1650.95}} = 0.962$$

$$\eta_1 = 0.6 \times 0.962 = 0.5772$$

η_2 is the allowable factor for combined action = 1.0 in the present case.

$$\sigma_{by} = \frac{M}{z}$$

$$z = \frac{5.63 \times 10^9}{(D/2)} = \frac{5.63 \times 10^9}{500} = 1.126 \times 10^7 \text{ mm}^3$$

$$\sigma_{by} = \frac{1 \times 10^9}{1.126 \times 10^7} = 88.81 \text{N/mm}^2$$

$\sigma_{bz} = 88.81$ N/mm^2 for a cylindrical member

Step # 7: To determine the critical bending strength, σ_{CBy}:
The critical bending strength depends on the following ratio:

$$\frac{D}{t} \leq \frac{E}{4.5\sigma_0}$$

$$\frac{1000}{15} = 66.67$$

$$\frac{E}{4.5\sigma_0} = \frac{2 \times 10^5}{4.5 \times 240} = 185.185$$

The above relation is satisfied; the critical bending strength is given by the following relationship:

$$\sigma_{Cb} = \begin{cases} \left(\dfrac{S_{MP}}{S_{ME}}\right)\sigma_0 \quad \text{if } \dfrac{\sigma_0 D}{Et} \leq 0.02 \\ \left[1.038 - 1.90\sigma_0\dfrac{D}{Et}\right]\left(\dfrac{S_{MP}}{S_{ME}}\right)\sigma_0 \quad \text{if } 0.02 \leq \dfrac{\sigma_0 D}{Et} \leq 0.10 \\ \left[0.921 - 0.73\sigma_0\dfrac{D}{Et}\right]\left(\dfrac{S_{MP}}{S_{ME}}\right)\sigma_0 \quad \text{if } \dfrac{\sigma_0 D}{Et} > 0.1 \end{cases}$$

Further, checking for the ratio

$$\frac{\sigma_0 D}{Et} = \frac{240 \times 1000}{2 \times 10^5 \times 15} = 0.08$$

S_{MP} and S_{ME} are plastic and elastic section moduli, respectively, and are given by

$$S_{MP} = \frac{1}{6}[D^3 - (D - 2t)^3] = \frac{1}{6}[1000^3 - (1000 - 2 \times 15)^3] = 1.455 \times 10^7 \text{ mm}^3$$

$$S_{ME} = \frac{I}{y} = \frac{5.63 \times 10^9}{500} = 1.126 \times 10^7 \text{ mm}^3$$

$$\sigma_{CBy} = [1.038 - (1.90 \times 0.08)]\left(\frac{1.455 \times 10^7}{1.126 \times 10^7}\right)240 = 274.77 \text{ N/mm}^2 = \sigma_{CBz}$$

Step # 8: Checking for the combined loading:

$$\frac{\sigma_A}{\eta_1 \sigma_{CA}} + \frac{1}{\eta_2}\left\{\left[\left(\frac{\sigma_{by}}{\sigma_{Cby}}\right)\right]^2 + \left[\left(\frac{\sigma_{bz}}{\sigma_{Cbz}}\right)\right]^2\right\}^{0.5} \leq 1.0$$

$$\frac{21.54}{0.5772 \times 231.63} + \frac{1}{1}\left\{\left[\left(\frac{88.81}{274.77}\right)\right]^2 + \left[\left(\frac{88.81}{274.77}\right)\right]^2\right\}^{0.5} = 0.618 < 1 \text{safe}$$

3 Fluid–Structure Interaction

3.1 INTRODUCTION

Fluid–structure interaction (FSI) plays an important role for structures placed in the path of a flowing fluid. The presence of the structure alters the fluid flow field in its vicinity. The degree of compliance offered by the structure adds further complexity due to the reaction it offers to the excited force. Even though the structures remain flexible (e.g., tension leg platform [TLP]) in certain degrees of freedom, dynamics become important. It is not their deformation capacity that is looked upon in this context. FSI becomes more important when the flow is steady, maybe in the form of a current or a vertical shear. However, in real sea states structures experience large oscillating forces in the flow direction. When structures are placed in the flowing fluid, the flow pattern is altered. Restraints are developed in the fluid medium to maintain the position of the structure. Forces acting on the structures in the fluid medium can be classified as drag (acting in-line with the direction of flow) and lift (transverse to the direction of flow). Further, drag force can be classified as higher- and smaller-frequency components. These components are functions of the geometry of the structure and flow conditions. Lift forces contain oscillatory components with multiple frequencies. On the downstream side, the flow will return to its unaltered condition due to fluid viscosity and damping. The region of the altered flow directly behind the structure is called the *wake region.* In the wake region there will be a one-to-one relationship between the extent of the wake region and the restraint loads. This implies that in the wake region frequency content is determinant and is the same as that of the restraint loads, which enables the determination of FSI in a closed form.

The data shown in Table 3.1 represent the case for a cylinder the axis of which is normal to the flow direction. Flow is without turbulence or boundary effects.

3.2 VERTICAL CYLINDERS IN UNIFORM FLOW

Vertical cylinders experience loading from the flowing fluid, and the FSI will be defined by the Reynolds number for cylinders that are infinitely long, smooth, and fixed. Cylinder roughness and fluid turbulence reduce boundary values on either side of the supercritical region. If the length of the cylinder is finite, this will introduce a 3D aspect to the flow. The ends of the cylinder will affect drag and lift coefficients; they are also dependent on the location of the cylinder. The vortex shedding pattern will also be affected (Table 3.2).

TABLE 3.1
Regimes in Uniform Flow

Flow Region	Reduced Velocity	Vortex Shedding	Types of Vibration Caused
I	1.7–2.3	Symmetric shedding	In-line oscillation only
II	2.8–3.2	Alternate shedding of vortices	Predominantly in-line vibrations Some transverse vibrations are also seen
III	4.5–8.0	Alternate shedding of vortices	Predominantly transverse vibration In-line vibrations are seen at frequency twice that of the transverse vibration This is called *figure eight motion*

TABLE 3.2
Reduced Velocity Range

Flow Region	Re Range	Flow Condition	Forces on Cylinder
Laminar	0–40	No separation of flow	Drag forces occurring in the direction of flow
Subcritical	40–5E05	Broken streamlines	Lift forces depend on the Strouhal number Steady drag force + smaller oscillating drag forces at double the frequency of lift force
Supercritical	5E05–7E05	Ill-defined vortices	Drag forces decrease rapidly Lift and drag forces will be seen at higher frequencies
Transcritical	>7E05	Vortices will be persistent Turbulent flow due to randomness in fluid viscosity	Similar to subcritical range

Source: Adapted from Humphries, J.A. and Walker, D.H. 1987. *Proc. Sixth OMAE*, Houston, TX, 2:139–143.

3.3 FLOW IN DEEP WATERS

In deep waters, flow is not uniform with depth and results in positive shear. Velocity will be greater than that near the surface. When a vertical cylinder is in a shear flow condition, a 3D flow regime will occur. Under positive shear, the wake region experiences vertical upward flow. Variation in stagnation pressure causes downward flow along the length. The flow is *sheared* from upstream to downstream. There is downward flow on the U/S side and upward flow on the D/S side, as shown in Figure 3.1.

The shear flow effect reduces the pressure coefficient at the top of the cylinder and increases the coefficient at the bottom. The pressure coefficient also changes

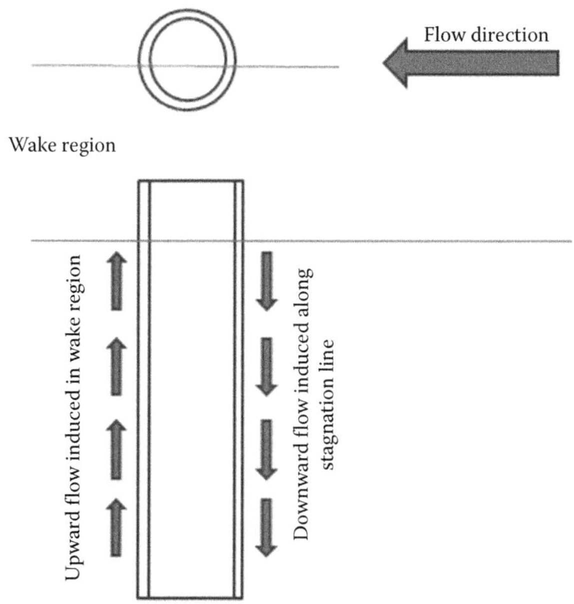

FIGURE 3.1 Flow in deep waters.

with the strength of shear. In uniform flow, vertical cylinder will show vortices at the same frequency over its entire length, whereas in shear flow, frequency changes continuously.

3.4 HORIZONTAL CYLINDER IN UNIFORM FLOW

Examples of horizontal cylinders are subsea pipelines, members of floating breakwaters, pontoons, and so on. These members will be influenced by current and waves. Under a uniform flow field, a horizontal cylinder will generate waves near the upper boundary (free surface). Wave resistance depends on the Froude number:

$$\text{Froude number } F_r = \frac{U_2}{(2gy_1)}$$

In the equation, y_1 is the depth of immersion, measured from the axis of the cylinder to the free surface. When $y_1 \gg$ radius, maximum wave resistance will occur, and the minimum will occur when $F_r < 0.375$. Boundary effects and end effects of the cylinder introduce a 3D effect in the flow past the cylinder of finite length. The major difference between horizontal and vertical cylinders is in the appearance of lift force. When the horizontal cylinder is located near the boundary, flow will become unsymmetrical. Lift force will become a function of cylinder diameter and distance to the boundary. Horizontal cylinders will also be subjected to flow-induced vibrations.

3.5 HORIZONTAL CYLINDER IN SHEAR FLOW

Under shear flow, velocity variation across the face of the structure will be significant. Ratio of turbulence to velocity variation across the horizontal cylinder is higher than that for a vertical cylinder. For an increase in shear parameter, the Strouhal number increases because of the increase in vortex shedding frequency. With a fixed horizontal cylinder, load at a given velocity for a positive shear is more than that for uniform flow. At the free surface, the vertical load in both positive and negative shear increases significantly in comparison to uniform flow. Dynamic vertical loads on the cylinder at mid-depth under positive and negative shear contain higher energy at high frequencies. Near the surface there is significant reduction. A horizontal cylinder at the free surface suppresses eddy shedding and wake formation. The steady component of the vertical load increases significantly for a horizontal cylinder at the free surface.

3.6 BLOCKAGE FACTOR

Closely spaced members, connected in different orientations, cause distortion in the fluid field around them. For closely spaced members, the structure becomes dense. For dense structures, the flow field slows down because it travels through the structure. This causes a blockage effect and complicates the actual velocity field around the structure. Load on the structure increases due to this blockage. Drag force is summed for each member in the dense structure. In the case of a group of vertical cylinders present in the flow field, blockage factor $CBF = 0.25S/D$ (for $0 < S/D < 4.0$) $= 1.0$ for $S/D = 4.0$, where S is the c/c distance of the cylinder and D is the diameter.

3.7 WAVE–STRUCTURE INTERACTION

When waves flow past a cylinder, they cause oscillating in-line forces on the cylinder. In addition, the free surface also changes in the case of a submerged cylinder. Large structures placed in a wave field alter incident waves in the vicinity. If the dimension of the structure is large compared with the wave length, the flow remains attached to the structure. When waves pass a cylinder, they cause an oscillating in-line force on the cylinder. In addition, the free surface also changes in the case of a submerged cylinder. Several procedures can be used to explain the potential function generated in the vicinity, knowing the incident wave potential. The flow around the structure is assumed to remain attached. Separation is neglected and the fluid is assumed to be incompressible and irrotational, and the structure is assumed to be rigid. Wave amplitude is assumed to be small. Fluid flow in the neighborhood is described by the velocity potential, which under linear theory, is given by

$$\Phi = \varphi e^{(-i\omega t)} \tag{3.1}$$

where φ is the spatial part of the total velocity potential and ω is the incident wave frequency. Total potential is the sum of the potential of the incident wave component and the potential of scattered wave component. The scattered wave component is normally represented by a continuous distribution of waves. It is assumed to be the superposition of numerous waves. The boundary value problem, defined in terms of Laplace partial differential equations, is transformed into original partial differential equations in potential theory. Boundary conditions include ocean floor, free and submerged surface of the structure, and radiation condition because flow approaches infinity.

3.8 PERFORATED CYLINDERS

3.8.1 Background Study

A detailed review of the literature on wave interaction on porous cylinders is presented. The influence of the porous region on the forces experienced by the inner cylinder in the presence of a perforated outer cover is highlighted. Both experimental and analytical studies carried out by various researchers show a common agreement of significant force reduction on the inner cylinder with a perforated outer cover. Several studies highlighting the dynamic response of offshore tension leg platforms under regular and random waves are also discussed, and various nonlinearities associated with the response behavior of TLPs under waves are presented. A few experimental investigations carried out on the dynamic response of TLP with perforated members are also discussed. It is seen from the above review that experimental investigations on dynamic response of TLPs with a perforated outer cover are scarce in the literature; hence, the current research study justifies the defined objectives.

3.8.2 Experimental Investigations on Perforated Cylinders

Perforated cylindrical structures reduce wave–structure interaction and scouring problems considerably. The existence of an exterior porous cylinder reduces the hydrodynamic forces on the inner cylinder caused by the direct wave impact. It is seen from the literature that for reduced annular spacing, long waves impose larger forces on the inner cylinder than short waves. On the basis of experimental investigations carried out, researchers recommended porosity of about 10%–15% beyond which no appreciable reduction in hydrodynamic response is seen. Preliminary experimental investigations are carried out to study the hydrodynamic response of perforated cylinders in regular waves. Variations of forces due to regular waves in a cylinder, with and without a perforated cover, are measured. The experimental setup for the study is shown in Figure 3.2. To evaluate the influence of porosity and the diameter of perforations, three scaled models comprising an outer cylinder of 315 mm diameter and an inner impermeable cylinder of 110 mm diameter were fabricated with uniform annular space. Steel frames were used for clamping the inner cylinder and the outer cylinder in the wave flume; the model was suspended with a clear gap of about 50 mm above the seafloor, ensuring a cantilever action. Strain gauges were placed along the inner cylinder to determine the forces. Regular waves of height ranging from 5 to 25 cm at time periods of 1–2 s were generated for the tests.

Details of the inner and outer cylinders used for the study are given in Table 3.3. The diameter and length of perforations were varied to achieve different porosity

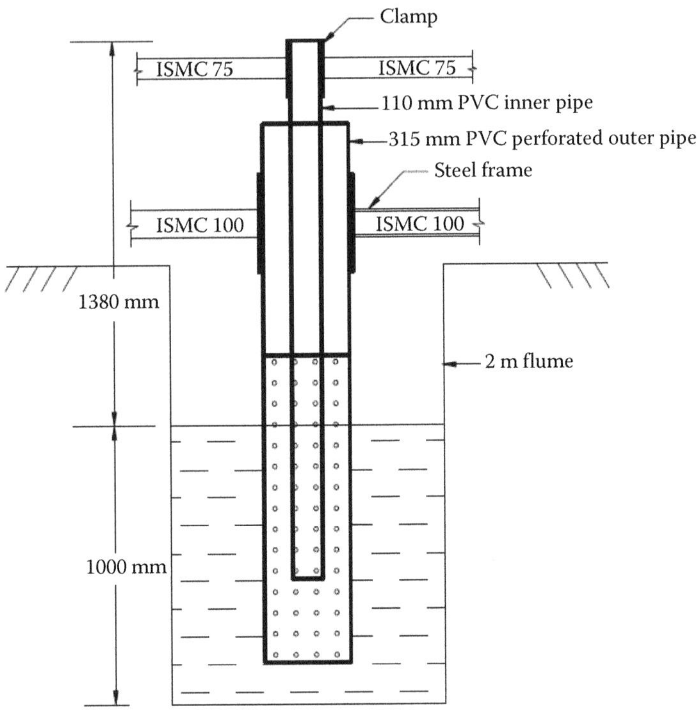

FIGURE 3.2 Experimental setup to study the response of a perforated cylinder.

ratios. Perforation ratios were in compliance with the Indian standard code of practice IS 4985:2000. Figure 3.3 shows the inner and outer cylinders with different perforations, and details of strain gauges affixed on the inner cylinder can also be seen.

The inner cylinder was fixed at one end and the other end was set free to enable the cylinder to behave similarly to the cantilever beam. Known bending stresses were

TABLE 3.3
Geometric Details of Cylinders Considered for the Study

		Outer Cylinder		
Description	**Inner Cylinder**	**A**	**B**	**C**
Diameter (mm)	110	315	315	315
Thickness (mm)	4.4	8.7	8.7	8.7
Perforation diameter (mm)	—	10	15	20
Length of the cylinder (mm)	1900	1930	1930	1930
Length of perforations (mm)	—	1450	1050	1050
No. of perforations along length	—	41	26	24
No. of perforations along the circumference	—	28	24	24
Porosity (%)	—	6.3	10.6	16

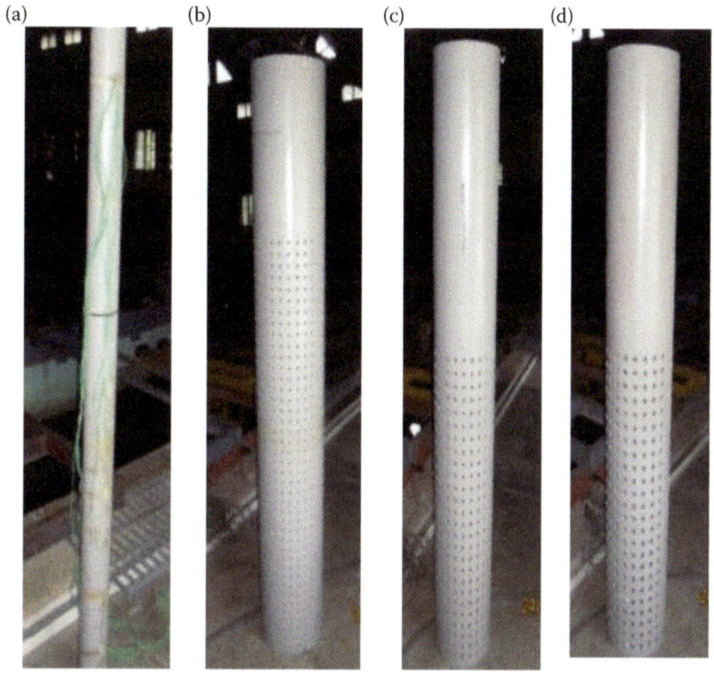

FIGURE 3.3 Perforated cylinders considered for the study: (a) inner cylinder; (b) outer cylinder; (c) outer cylinder; (d) outer cylinder.

created by applying the point load at a constant distance of 50 mm from the free end. Bending strain in the inner cylinder, with and without perforated outer cylinders, was measured during the passage of regular waves. Regular waves with wave heights of 5–25 cm in the intervals of 5 cm and wave periods of 1–2 s at intervals of 0.2 s were considered in the study. Bending strains were postprocessed to determine the hydrodynamic forces on the inner cylinder; their variations along its length were also studied. Maximum values of hydrodynamic forces computed on the inner cylinder encompassed by outer cylinders with different porosity were measured; a typical value for 15 cm wave height is given in Table 3.4. It is seen that force reduction decreases with increases in porosity because the inner cylinder shall be exposed to more hydrodynamic load due to increased porosity. Further, the force on the inner cylinder decreases significantly for short period waves compared with long period waves.

Figures 3.4 and 3.5 show the force variation on the inner cylinder encompassed with outer cylinders with different porosities; the plots are drawn for different wave heights varying from 5 and 25 cm, respectively. Wave periods were selected appropriately with respect to the cylinder diameter so that the model remains in the Morison regime. It is seen from the figures that force variation in the inner cylinder is nonlinear; decrease in wave force is not proportional to either increase in time period or wave height.

The model investigated was built to a scale of 1:140, and the force reduction corresponding to the prototype cylinders with different porosities is shown in Table 3.5.

TABLE 3.4
Hydrodynamic Forces for 15 cm Wave Height (N)

		Outer Cylinder		
Wave Period (s)	**Inner Cylinder**	**A**	**B**	**C**
1	15.38	5.18	9.49	10.12
1.2	14.35	4.80	6.77	9.50
1.4	11.49	4.50	6.18	7.69
1.6	10.54	3.45	4.75	6.05
1.8	9.59	3.20	4.75	6.48
2	8.75	3.63	4.50	5.82

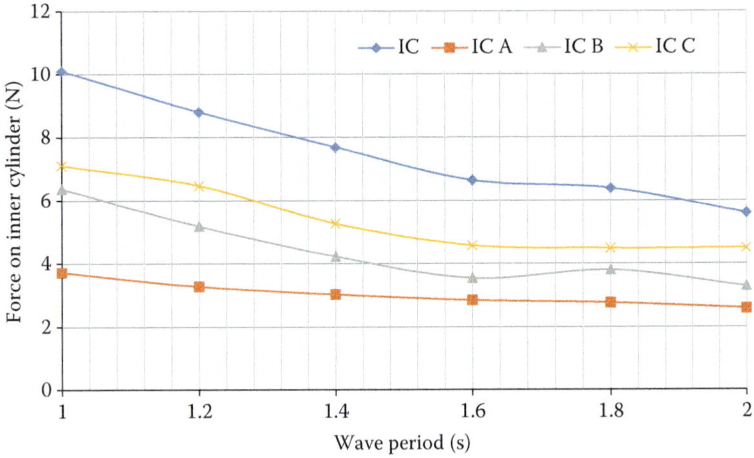

FIGURE 3.4 Force variation in cylinders ($WH = 10$ cm).

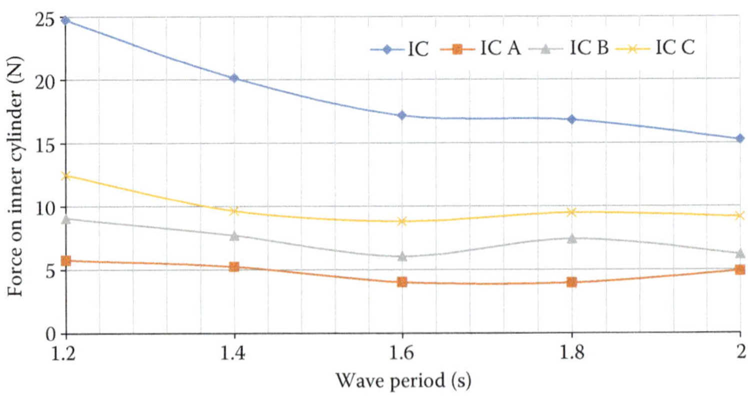

FIGURE 3.5 Force variation in cylinders ($WH = 25$ cm).

TABLE 3.5
Force Reduction in Inner Cylinder

Sl. No.	Description	Model (1:140)	Prototype
1	Water depth (m)	1.0	140
2	Diameter (inner cylinder)	0.11 m	15.4 m
	Force Reduction ($H = 25$ cm, $T = 1.2$ s)		
3	With outer cylinder A	18.97 N (76.59%)	52.05 MN
	With outer cylinder B	15.70 N (63.38%)	43.08 MN
	With outer cylinder C	12.24 N (59.63%)	33.58 MN
	Force Reduction ($H = 5$ cm, $T = 2$ s)		
4	With outer cylinder A	1.37 N (35.54%)	4.62 MN
	With outer cylinder B	0.85 N (29.02%)	2.86 MN
	With outer cylinder C	0.51 N (17.41%)	1.72 MN

On the basis of experiments conducted, it is seen that the hydrodynamic forces on the inner cylinder decreases with the decrease in porosity. This reduction in forces is significant for short period waves when compared with long period waves; variation is nonlinear and is not proportional to either the increase in wave height or wave period. Size of perforations and porosity influences hydrodynamic forces on cylindrical members significantly. It is also seen that the maximum force reduction is about 76% and the minimum is about 17%. The present study highlights the advantages of deploying perforated cylinders as outer covers on impermeable inner cylinders. This has direct application of force reduction on the inner cylinders; although the application is not required for new designs, it is recommended as an alternative method for retrofitting of offshore structures. The derived conclusions, based on the experiments conducted, emphasize the known concept of force reduction on inner cylinders; the study quantifies the values for different porosities and size of perforations, which can be seen as an original contribution to the design of offshore structures with perforated members.

3.8.3 Experimental Investigations on Perforated TLP Model

The pilot study of this project was focused on detailed experimental investigations carried out on an impermeable inner cylinder encompassed by a larger outer cylinder with perforations along its length. By varying the porosity and diameter of perforations, their influence on the hydrodynamic response of the cylindrical member was examined. In an extended concept of the study, offshore TLPs with perforated members were experimentally investigated.

Offshore TLPs are hybrid structures whose heave motion is highly damped, giving operational advantages; however, large surge, sway, and yaw motions cause inconvenience to people on board although the platform remains stable for operational sea state. To reduce the wave impact on pontoons and cylindrical members of TLPs, perforated cylinders can be used as outer covers to the members at highly stressed regions. This is one of the practical techniques to retrofit coastal and offshore

TABLE 3.6
Average Surge Response Reduction

Wave Period (s)	Average Response Reduction
0.80	13.35%
1.20	7.07%
1.60	6.78%
2.00	18.01%
2.40	24.84%

structures and also to improve their structural safety. In this experimental study, a 1:150 scale model of a prototype TLP was fabricated with and without perforated outer columns. Dynamic response in various active degrees of freedom and tether tension variations were studied under regular waves of different time periods and wave heights. Details of the model are given in Table 3.6. Figures 3.6 and 3.7 show the views of the TLP model considered for the study.

The outer perforated column was 150 mm in diameter and of height 180 mm. The middle third portion of the outer column was perforated with holes of 5 mm diameter spaced at 12 mm intervals, resulting in a porosity of 13.5%; the outer column was connected to the inner column through a 10 mm thick ring to maintain the required annular space. A reasonable comparison was possible through the attempted study because the static characteristics like mass and the centre of gravity (KG) of the model remain nearly the same irrespective of the presence of the perforated cover. Figure 3.8 shows the experimental setup of the study.

Free-vibration tests were carried out to determine the natural frequency and damping coefficient of the model. The model was subjected to waves in the head sea direction whose time periods were varied in the range of 0.8–2.4 s; wave heights were varied in the range of 5–9 cm. Figures 3.9 through 3.11 show the response of the TLP model in surge and heave degrees of freedom and tether tension variation, respectively.

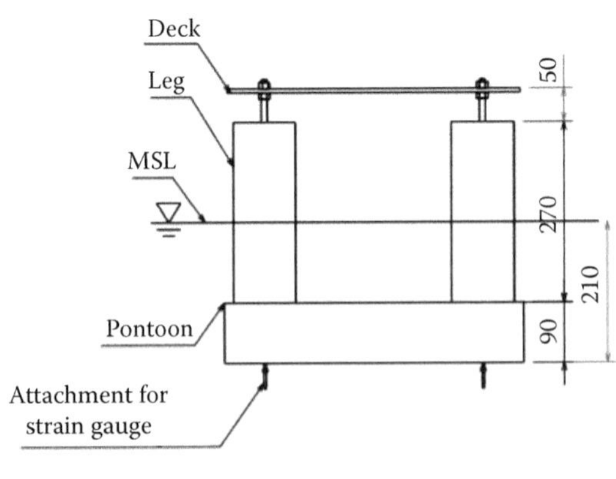

FIGURE 3.6 Front view of TLP model without the perforated cover.

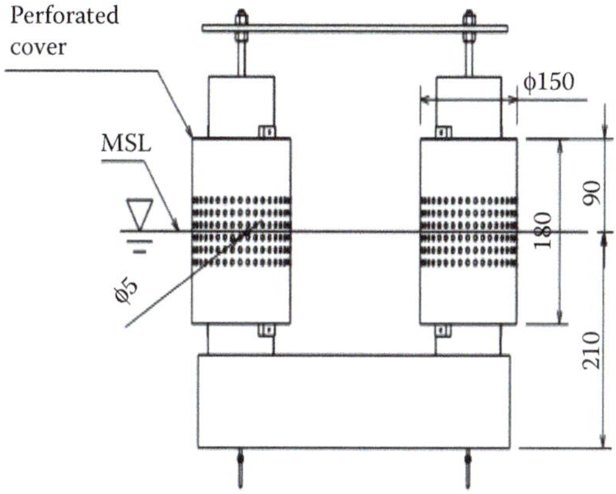

FIGURE 3.7 Front view of TLP model with the perforated cover.

It is seen that the response of TLP is reduced in the presence of the outer perforated cover. Tether tension variation also shows a significant reduction in the presence of the outer perforated cover. Hydrodynamic responses of TLPs with outer perforated covers are also investigated for different wave approach angles. Table 3.6 shows the average reduction in surge response for different wave periods; it is seen that the maximum response reduction is about 24% due to the presence of outer perforated cylinders.

This study highlights the advantages of deploying perforated column members for TLP because outer covers on impermeable inner cylinders. This has direct application of force reduction on the inner cylinders; though the application is not required for new designs, it is recommended as an alternative method of retrofitting of offshore structures.

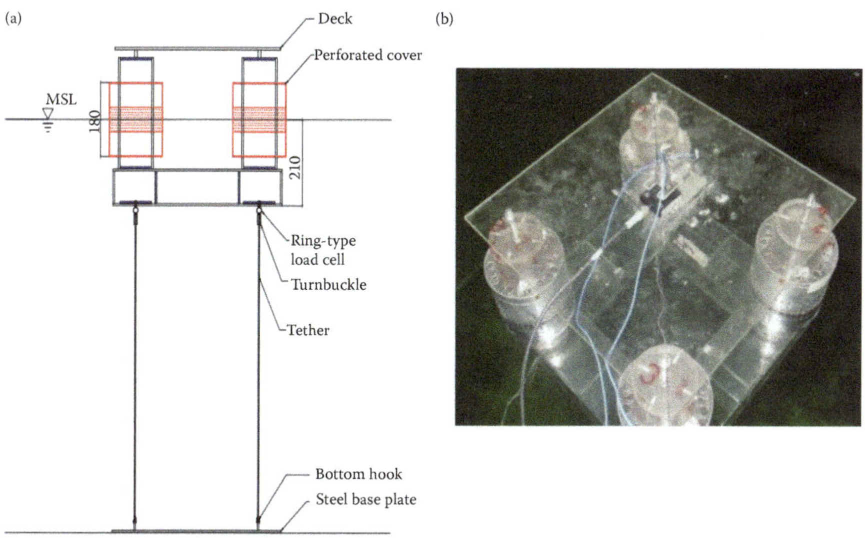

FIGURE 3.8 Experimental setup: (a) components of the model and (b) instrumentation.

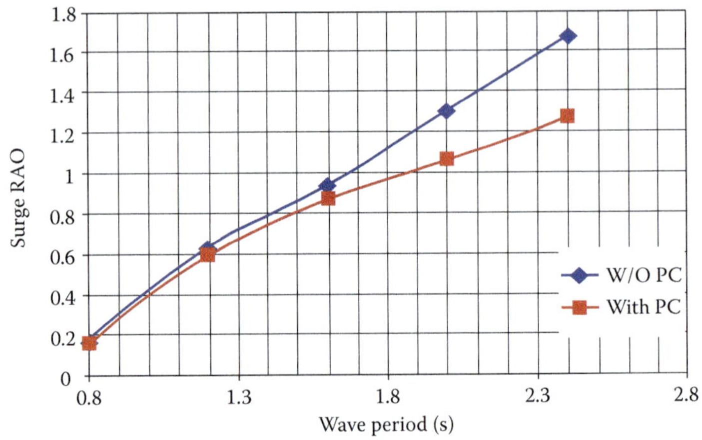

FIGURE 3.9 Surge RAO for 7 cm wave.

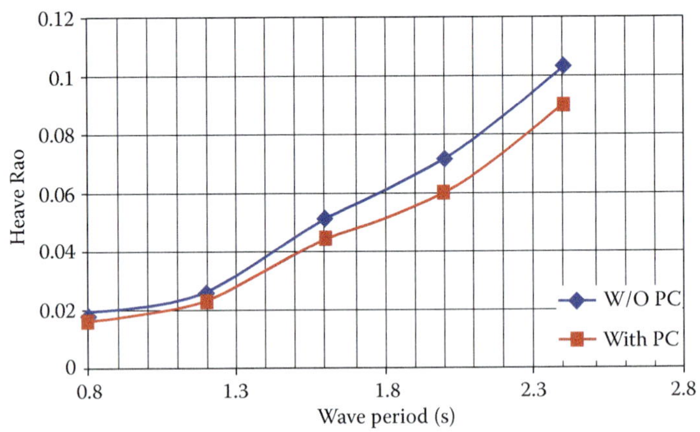

FIGURE 3.10 Heave RAO for 7 cm wave.

3.8.4 Numerical Studies on Perforated Cylinders

Numerical studies on perforated cylinders are carried out through an application of the simulation software STAR-CCM+. An attempt is made to simulate the hydrodynamic response of the perforated cylinder with a porosity of 6.3%, which is similar to that of the perforated outer cylinder designated as A in the experimental studies. Simulation through STAR-CCM+ software is chosen due to the numerical capabilities enabled in different modules to simulate viscous drag and turbulence effects that are caused by perforations. Details of the simulation, as attempted through several stages of the numerical modeling, are discussed in steps; various settings such as mesh and physics models used in the study and their significance are also presented. A model of the perforated cylinder in CATIA V5 is shown in Figure 3.12.

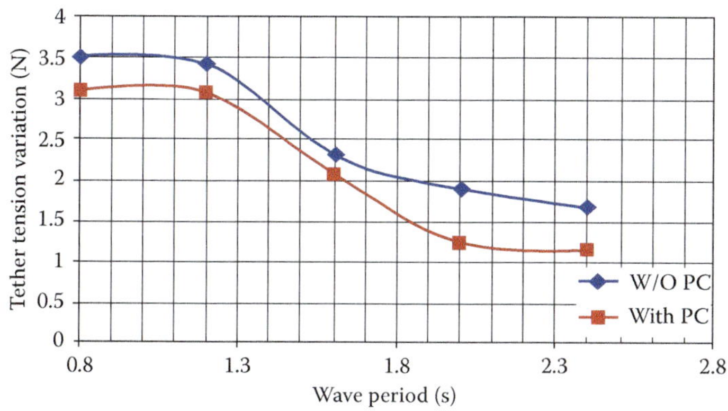

FIGURE 3.11 Tether tension variation for 7 cm wave.

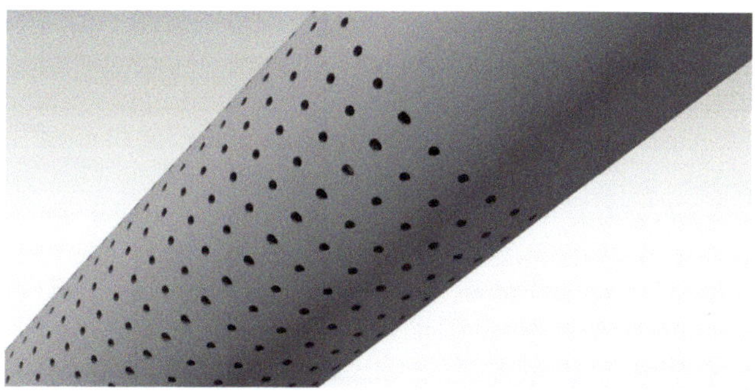

FIGURE 3.12 Perforated outer cylinder.

Two different geometries, namely: (i) inner cylinder and (ii) inner cylinder with perforated outer cylinder were exported from CATIA V5 and new simulations were generated in STAR-CCM+ from the imported files. Figures 3.13 and 3.14 show the mesh generation of the inner cylinder and inner cylinder with perforated outer cover, respectively; distribution of elements attained using the chosen volume control makes the mesh denser in the fluid region, which can be seen from the figures.

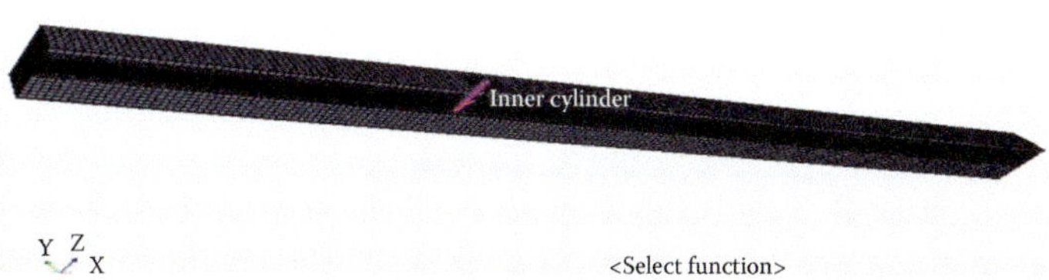

FIGURE 3.13 Domain of inner cylinder generated with volumetric control.

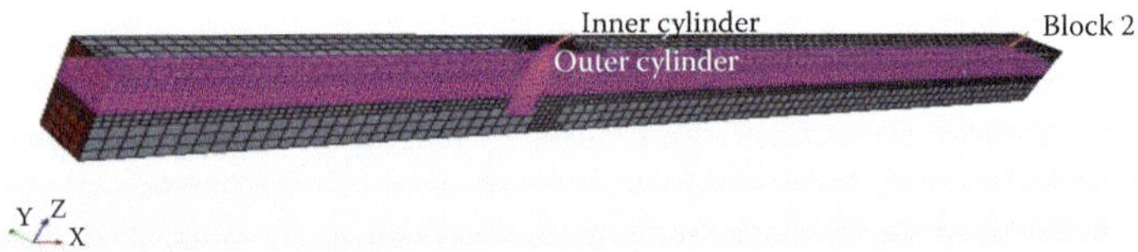

FIGURE 3.14 Domain of inner cylinder with perforated outer cylinder.

The domain of inner cylinder with perforated outer cylinder was generated with volumetric control. Subsequently, the "generate volume mesh" tool was used to generate the mesh; 716,801 cells and 2,167,056 faces were generated for simulation of the inner cylinder while 3,242,875 cells and 9,671,484 faces were generated for that of the inner cylinder with a perforated outer cylinder, which is fairly a dense mesh. Several physics models were activated to simulate the wave forces on both the numerical models of an inner cylinder and an inner cylinder with a perforated outer cylinder. A total of 16 physics models were used in the present simulation and activated. A new first-order wave was created under the "waves" child node of the "VOF waves" node in the list of chosen physics models. The "point on water level" was set to 0.54 m for the simulation of inner cylinder and 0.8825 m for the simulation of inner cylinder with a perforated outer cylinder; the chosen values also match the relevant values used in the experimental investigations. Wave amplitude was set to be 0.05 m. Specification type was set to "wave period" and numerical simulations were run for six waves for each model, for wave periods of 1.0–2 s with an interval of 0.2 s. The "volume fraction" was set to composite and the method of each of the phases, water, and air set to "field function." The boundaries in the region were set to match various types of boundary conditions, namely: (i) the "inlet" boundary was set as a velocity inlet; (ii) "outlet" boundary as a pressure outlet; and (iii) "inner cylinder" and "the outer perforated cylinder" were set as wall boundaries. The "velocity specification" method in the velocity inlet was changed to "components" and the "velocity" and "volume fraction" values were set identical to match the experimental setup. Similarly, the "volume fraction" and "pressure" settings of the pressure outlet were also set accordingly. The "time-step" property of the "implicit unsteady" solver was set to 0.01 s. Under the "stopping criteria," the "maximum inner iterations" property was set to 10 and the "maximum physical time" was set to 10 s. The "initialize solution" tool was selected to activate the required simulation. An isosurface with an isovalue of 0.5 and a scalar set to "volume fraction > water" was used to visualize the free surface. A scalar was used to visualize both the inner cylinder and inner cylinder with the perforated outer cylinder.

Both the simulated models, namely inner cylinder and inner cylinder with the perforated outer cylinder, were subjected to unidirectional waves of 10 cm wave height. Wave periods were varied from 1 s to 2 s with 0.2 s variation; the Reynolds-averaged Navier–Stokes equation was solved, which was assumed to be converged when the residuals

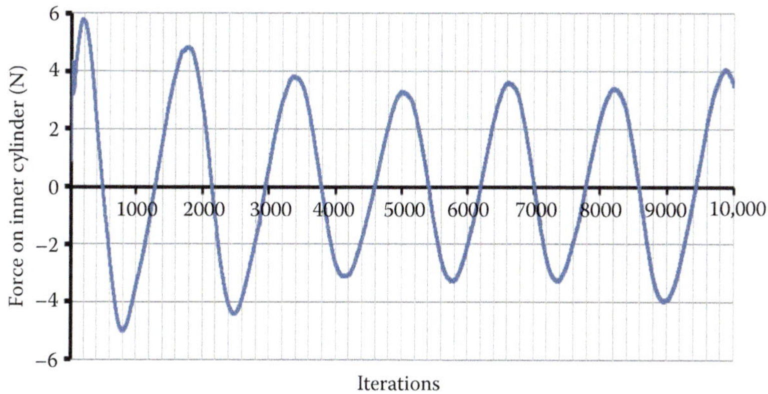

FIGURE 3.15 Force on inner cylinder ($WH = 10$ cm; $WP = 1.6$ s) in numerical simulation.

decreased by multiple orders before settling around 0.001. Figures 3.15 and 3.16 show the variation of forces on the inner cylinder with and without the perforated outer cover, respectively, for wave height of 10 cm and wave periods ranging from 1 to 2 s.

It was observed that forces on the inner cylinder without the perforated outer cover, computed from both the numerical simulations and experimental results, range from 1.99% to 13.63% with an average error of 7.08%. In the case of forces computed on the inner cylinder with the perforated outer cylinder, errors between numerical simulations and experimental results ranged from 2.51% to 11.28% with an average of 7.5%. Figure 3.17 shows a graphical comparison of results obtained from numerical simulation and experimental investigations. It is also seen from the figure that both the results agree well within the acceptable error of tolerance for the chosen range of wave periods.

On the basis of experimental investigations and numerical studies carried out, it is seen that the force estimates between both the studies are in good agreement,

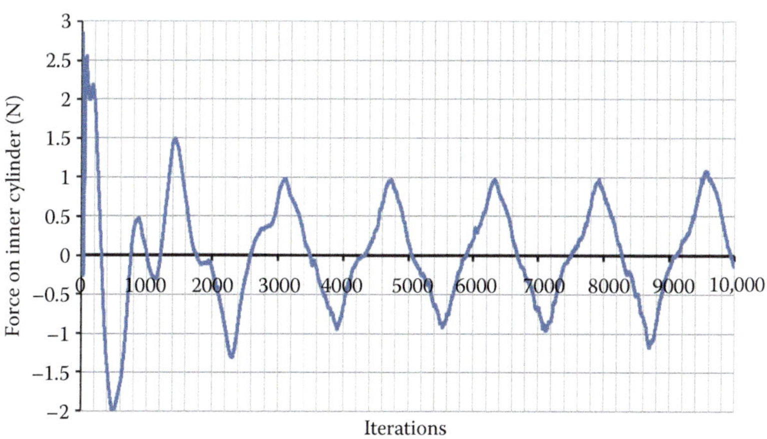

FIGURE 3.16 Force on inner cylinder with perforated outer cylinder in numerical simulation ($WH = 10$ cm; $WP = 1.6$ s).

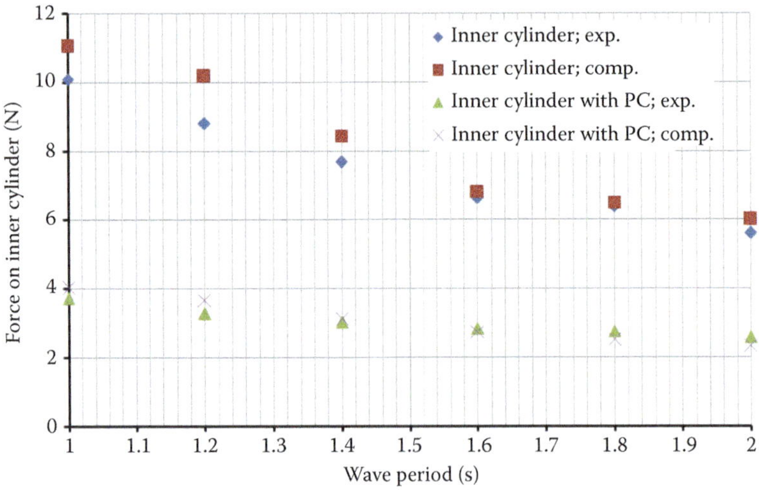

FIGURE 3.17 Comparison of forces on the inner cylinder with and without perforated outer cylinder.

validating the numerical procedure adopted in the study. Few problems associated with the simulation, namely: (i) high cell numbers and (ii) larger domain attempted to simulate the conditions because that of the wave flume resulted in not yielding the velocity profile variation at the desired points along the cylinder. An attempt will be made to simulate the numerical model for a 2D plate with a relatively smaller domain so as to trace the variation of velocity profile along the water depth; in addition, parametric studies such as (i) size of perforation; (ii) perforation ratio; and (iii) location of perforation will be examined to derive the velocity profile variation under the influence of the chosen parameters. When the train of waves hits the cylinder the energy get dissipated due to the back and forth movement, causing a partial breaking of waves. Hence this phenomenon acts as a good energy dissipation technique which is an economical approach preferred by design engineers.

3.9 EVALUATION OF FORCE ON THE PERFORATED MEMBER

The forces on the inner cylinder are evaluated numerically using the classical approach using the Morison equation. The variations of water particle kinematics along the depth of the cylinder are derived from the numerical analysis for different chosen geometric configurations. The variations of water particle kinematics are presented in the form of design charts. Equation 6.1 gives the Morison equation used to evaluate the forces on the member. Both drag and the inertial components are functions of water particle kinematics:

$$dF = \left(C_m \rho \frac{\pi D^2}{4} \dot{u} + \frac{1}{2} C_D \rho D |u| u \right) dz$$

where C_m is the inertia coefficient, D the diameter of the cylinder, $\dot{u}$ the water particle acceleration, ρ the mass density of fluid, C_D the drag coefficient, and u is the instantaneous water particle velocity. For this study, evaluation of forces on the member is considered in accordance with rough cylindrical value, and velocity variation along the depth of the cylinder is obtained from the design charts. The variations of water particle kinematics are obtained in the form of design charts considering all possible parameters.

The developed charts that quantify the velocity variations along the water depth for different sea states chosen for the study can be seen as design aids to handle this complexity. Computed forces using the design charts on the inner cylinder are shown in Table 3.7.

Figure 3.18 shows the variation of the forces on the inner cylinder with the perforated outer cover obtained using numerical calculation and from design charts for different wave amplitudes ranging from 5 to 25 cm with an interval of 5 cm, similar

TABLE 3.7
Average Force Reduction in Inner Cylinder

Wave Period (s)	$H = 5$ cm	$H = 10$ cm	$H = 15$ cm	$H = 20$ cm	$H = 25$ cm
1	3.89 N	7.25 N	11.68 N	—	—
1.2	3.62 N	5.96 N	8.46 N	10.05 N	11.58 N
1.4	2.75 N	4.74 N	7.44 N	9.57 N	9.98 N
1.6	2.15 N	3.79 N	5.37 N	6.96 N	7.54 N
1.8	2.04 N	4.04 N	5.31 N	7.09 N	8.64 N
2	2.30 N	3.66 N	5.36 N	6.16 N	7.14 N

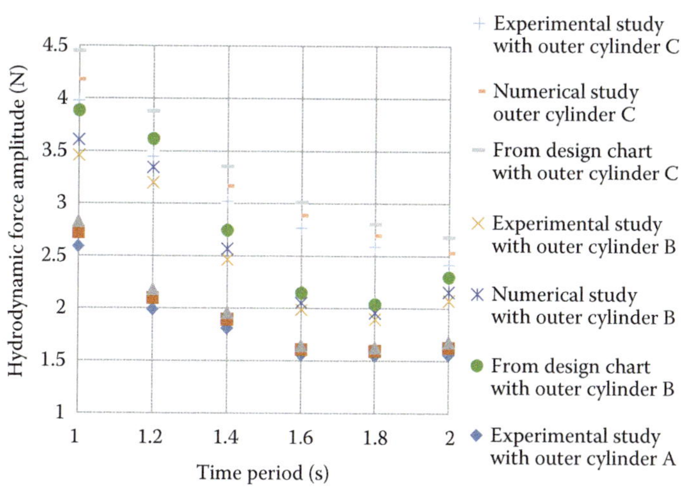

FIGURE 3.18 Variation of force on the inner cylinder with and without perforated outer cover ($H = 5$ cm).

to preliminary experimental investigation. The result shows a very good agreement for the longer periods and low steep waves.

3.10 ESTIMATION OF FORCE REDUCTION OF THE ELLIPTICAL MEMBER

A simulation is made for the elliptical member with and without perforation around the structure. The surface mesh is generated using SOLIDWORKS as shown in Figure 3.19a and b. Figures 3.20 and 3.21 show the simulation model for the elliptical member without and with perforation respectively.

The force on the member is quantified numerically using STARCCM+. The forces are evaluated for the model without perforation as 0.14544 N and with

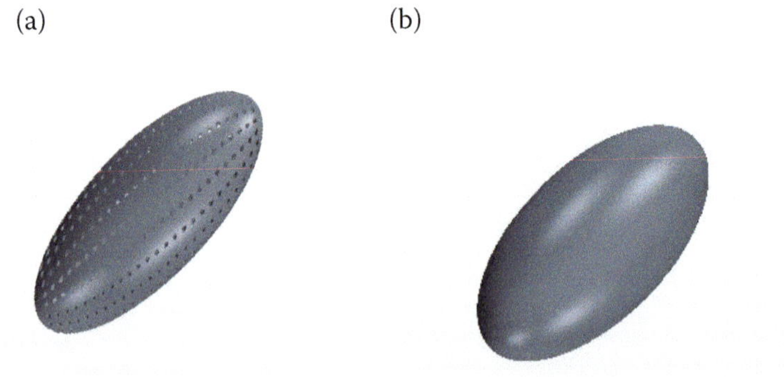

FIGURE 3.19 Elliptical member. (a) With perforation and (b) without perforation.

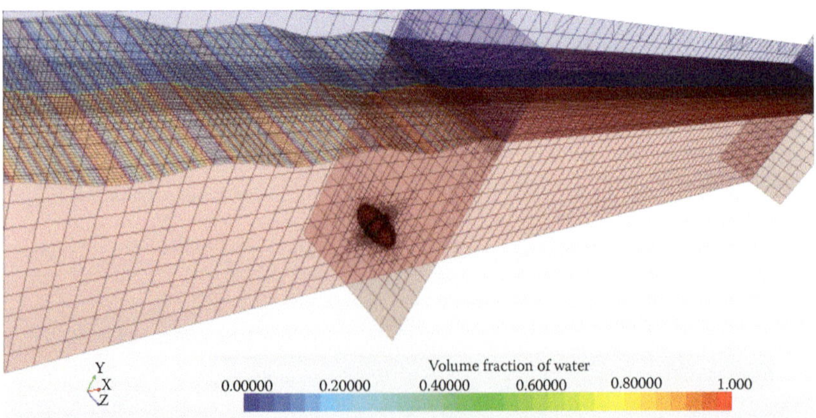

FIGURE 3.20 Numerical model of elliptical member without perforation.

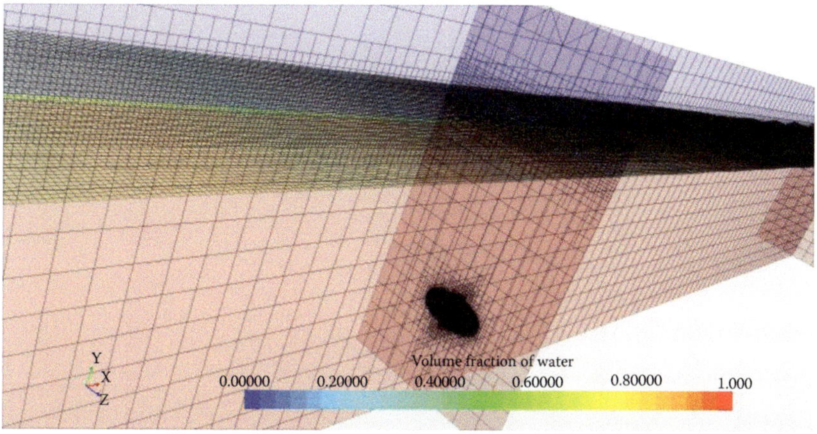

FIGURE 3.21 Numerical model of elliptical member with perforation.

perforation as 0.13851 N in model scale. The force reduction is quantified as only 5%. This method is not very significant as the horizontal velocity is almost nil at the specified level, and hence the mechanism of force reduction is not very significant. The retrofitting technique is more effective near the free surface where the horizontal velocity is maximum.

4 Reliability of Marine Structures

4.1 INTRODUCTION

Environmental loads that act on marine structures are not completely deterministic; only a few features of the loads are known (Ang and Tang, 1975a). Some of the examples of such loading are extreme waves, sea bed movement, wind loads on super structure, ice loads, impact loads caused by vessels, etc. Apart from these loads being random in nature, they also have a high probability of exceeding the safe design limits, which can cause an overloading situation (Benjamin and Cornell, 1970). In addition to the above, at all stages of construction of marine structures, there is high degree of uncertainty with regard to geotechnical problems that are site-specific (Chandrasekaran, 2013a,c; Chakrabarti, 1987, 1990). This shall transfer the complexities to a different level that uses a set of expressions to obtain the desired design quantities such as ultimate load strength, capacity of joints under different load combinations, additional stresses induced due to corrosion, etc.

The failure of engineering systems is a consequence of decisions made under uncertain conditions (Chandrasekaran and Bhattacharyya, 2012). It is therefore case-specific and cannot be uniformly applied to all engineering systems, in general. Reliability, therefore, implies estimate of limit state probabilities of the marine structure under adverse environmental loading, while safety is used to indicate reliability.

4.2 SAFETY AND RELIABILITY

Safety is related to an existing process. It has direct consequences to failure. It is a deterministic approach. It is a not a design method but useful only for assessment. Safety assessment methods give a close-form solution and therefore shall be applied to an existing structural system to assess its performance. On the other hand, *reliability* is a probability of the realization of safety (Benjamin and Cornel, 1970). It has a converse consequence of failure. It is a probabilistic approach and therefore used as one of the design methods (Ang and Tang, 1975a,b). Accuracy of the results of a reliability approach essentially depends on the data from which the results are arrived. Therefore, this has a probability of giving erroneous results if the data used are not reliable. Most importantly, reliability is assessed even before failure is foreseen and therefore reliability methods are accorded as a preventive forecast of failure (Chandrasekaran and Saha, 2011). While safety is based on statistical judgment, reliability methods are based on engineering judgment. Hence, experience plays an important role in the accuracy of results of reliability methods. *Risk* is an extension of reliability, which also includes the consequences of failure.

As both safety and reliability are circumscribed around failure probability, the definition of failure becomes important (Madsen et al., 2006). It is generally expressed in probabilistic terms and is assessed by the inability of a system to perform its intended function adequately on demand for a period under specific conditions (Daniels, 1945). The converse of the probability of failure is reliability, which is defined in terms of the success of a system. Reliability is the probability of a system performing its required function adequately for a specified period under stated conditions. The most important aspect of reliability is accounting for the uncertainties that make marine structures vulnerable to failure for a predefined limit state. *Accuracy of reliability analysis depends on how accurately all the uncertainties are accounted for in the analysis.*

It is an extremely complex and difficult task to consider all factors related to reliability of marine structures. This is not only due to the fact that marine structures are supposed to remain safe during their service life, but also due to lack of information on relevant factors that challenge their safety. It is important to note that safety requirements of marine structures also include accidents that are caused by ship collision, blowouts, and fire, which are not caused by failure in the structure; these factors comprise another level of uncertainties (Furnes, 1977). Unfortunately, reliable information on such accidental loads that challenge the safety of marine structures is not available; designers are forced to extrapolate information from other similar incidents (Chandrasekaran, 2013c; Nordic Committee for Building Regulations, 1977). Therefore, a great degree of uncertainty is introduced in the design stage itself, which adds to the complexity of reliability analysis because this actually does not arise from the failure of the structure.

Uncertainties that arise from the material properties comprise another set of important domain of interest (Ditlevsen, 1977; Naes and Moan, 2013). As discussed earlier, many discrepancies in ascertaining failure load based on the type of failure under a given triaxial stress state exist. Under such circumstances, a special caution is to be taken with respect to fatigue. These uncertainties clearly indicate that safety assessment of marine structures shall be considered from a probabilistic point of view. Therefore, reliability is expressed in probabilistic terms. A quality of performance is expected over a period under certain specified conditions. It is important to note that the expected quality of performance is only for a predefined period and not for infinite time. Therefore, in the reliability estimates as applied to marine structures, deterioration that arises from the environmental conditions are not accurately assessed (Fjeld, 1977).

4.3 INACCURACY IN RELIABILITY ESTIMATES

Reliability estimates, as applied to marine structures, cannot be accurate because of the following:

- It is practically impossible to identify all uncertainties present in the marine environment (Marshall, 1969, 1976). Only important uncertainties are identified and engineering judgment is applied to account for better accuracy.

- Methods of modeling and analysis are highly complex. This is not only due to complications that arise from the interaction of various engineering systems in marine structures but also from environmental factors (Høeg and Tong, 1977). Hence, some degree of uncertainties are always associated with modeling and analyses methods, which are an important aspect in the reliability framework.
- Analytical formulation of limit state surface and integration of probability density function within the domain of interest is very complex. This results in various approximations, which are subsequently described (Marshall, 1969).

Due to the above facts, different degrees of simplifications are made in the reliability estimates of marine structures. It is not possible to obtain an exact probability of failure of any structural system, except for a very simple one. Marine structures are highly complex due to their integral functionality and hence the inaccuracy.

Reliability methods are very important in marine structures because probabilistic-based design is helpful in estimating the degree of uncertainties with better accuracy. It is therefore helpful in optimizing the design process. In case of a well-defined rational analysis, assumptions and idealizations that are another form of uncertainties still exist. *Failure of a structural system* is everyone's interest under the life cycle assessment of marine structures. Failure is generally expressed as probability of failure. It is assessed by the inability of the structure to perform its intended function adequately on demand for a period under specific conditions. The period of time, during which the structure is unable to perform, is called *downtime or shut down time.* Reliability, the converse of probability of failure, is expressed in terms of success of a system. It is the probability of a system to successfully perform its intended function adequately for a specified period under stated conditions (Melchers, 1999). It is important to note that reliability estimates guarantee satisfactory performance of the structure only under certain (preset) conditions (Meyerhof, 1976; Moan and Sigbjørnson, 1977).

4.4 UNCERTAINTIES IN MARINE STRUCTURES

In dealing with the design of marine structures, uncertainties are unavoidable. Uncertainties are broadly classified into two types: (i) those associated with normal randomness and (ii) those associated with erroneous predictions and estimations of reality (Moses, 1977, 1976). The former type is called the *aletory* type whereas the latter is called the *epistemic* type. Irrespective of the classification, understanding the nature of their randomness is necessary. The aletory type generally arises from the loads that result from nature (e.g., earthquakes and floods). On the other hand, the *epistemic* type needs to be reduced using appropriate prediction models and sampling techniques. Uncertainties that arise from randomness and variability of excitation are irreducible, while statistical uncertainties that arise from estimation of parameters describing statistical models are reducible. While these uncertainties can be effectively handled by *Bayesian approach*, those arising from imperfections in mathematical modeling need to be minimized by following appropriate optimization procedures.

4.4.1 Bayesian Approach

In this approach, models and model parameters are developed by likelihood functions. With the help of these functions, posterior parameters or models are obtained from the prior ones. The posterior ones thus developed are supposed to have fewer uncertainties for the simple reason that their development is based on more data and observations. It is necessary that the designers are aware of many assumptions and idealizations on which methods of analysis and design are developed. Use of analytical tools shall be combined with sound engineering judgment based on experience and observations. In the last two decades, the need for solving complex problems has led to the development and use of advanced quantitative methods of modeling and analysis. However, it is a fact that the information derived from such sophisticated methods of analysis will be useful only if comprehensive input data are available and they are reliable too. Thus, the question of uncertainty and randomness of data is central to design and analysis. Probabilistic methods enable a logical analysis of uncertainties and provide quantitative basis for assessing the reliability of structures; the key requirement is the necessity of good practical experience to apply appropriate engineering judgment at various stages of analysis and design. Practical experience is always important and the observational approach can prove to be valuable; yet, the capacity to benefit from these is greatly enhanced by the rational analysis of uncertainty.

4.5 TYPES OF UNCERTAINTIES

The first group consists of uncertainties in material parameters, such as the modulus of elasticity of concrete and steel under different loading conditions. The second group consists of uncertainties in loads. Under static loading conditions, we are concerned with dead and live loads, and there are usually more uncertainties with respect to live loads. In particular, dynamic loads that are random in nature will have significant uncertainties. It is interesting to note that uncertainties associated with dynamic loads are very high in magnitude when compared with those arising from materials. Often, uncertainties associated with static loads may be negligible in comparison to those associated with material parameters. This is due to the strict quality control exercised in the manufacturing processes of construction materials used for marine structures. Nevertheless, it is interesting to know that material characteristics change under dynamic loads. One major variation can arise from the dynamic modulus of elasticity (E_D). A third group is related to the uncertainties that arise from mathematical modeling and methods of analysis. Each method of analysis or design is based on a number of assumptions, which generally account for the inherent uncertainties. Design principles address uncertainties by way of arbitrary safety factors (Moses and Stevenson, 1970). In fact, it is important to know the basis of these safety factors and the levels of uncertainties they address instead of relying on them blindly.

4.5.1 Dynamic Modulus of Elasticity

This is the ratio of stress and strain under vibrating conditions of the member (material) as computed from either the free or forced vibration tests in shear, compression, or elongation. It is an important characteristic of viscoelastic materials. In case of pure elastic materials, stress–strain variation occurs in phase so that the response to loads occurs simultaneously with the application of load; strain caused in the material is almost instantaneous and in phase with that of application of stress. On the other hand, in the case of viscoelastic materials, stress–strain behavior shows a phase lag; strain usually lags stress by ($\pi/2$) radians. Hence, the following relationship holds good:

$$\epsilon = \epsilon_0 \sin(\omega t) \tag{4.1}$$

$$\sigma = \sigma_0 \sin(\omega t + \varphi) \tag{4.2}$$

While elastic modulus describes the stress–strain behavior under monotonic loads, dynamic modulus describes the behavior under cyclic or vibratory loads. The following factors affect the dynamic modulus of elasticity (John et al., 2008):

- Dependence on the specimen type and experimental setup. Specimens with nonstandard shape/size and the variation in rate of loading, type of loading, etc. can affect the results significantly.
- Dynamic modulus of concrete obtained by different dynamic tests do not agree.
- Dynamic modulus obtained from pulse-wave propagation measurements are very high in comparison to those obtained from vibration measurements, even at resonance, carried on the same specimen.
- Dynamic modulus computed from specimens of prismatic sections are different from that of cylinders for the same concrete mix; this indicates the dependence of dynamic modulus on the shape and size of the specimens.

4.5.1.1 Longitudinal Vibration Frequency Method

Dynamic modulus is determined using longitudinal vibration frequency. The procedure assumes the following: (i) the analysis is effective for one-dimensional behavior and (ii) only plane-section motion is considered. Dynamic modulus can be computed by (ASTM C215)

$$E_D = \frac{v_P^2 \rho(1+\mu)(1-2\mu)}{(1-\mu)} \tag{4.3}$$

where V_p is the P-wave velocity, μ is Poisson's ratio, and ρ is the mass density.

4.5.1.2 Two-Frequency Method

Dynamic modulus can be determined using the two-frequency method as proposed by Love (1994) and Subramaniam et al. (2000). The expression proposed by the researchers yields dynamic Poisson's ratio, which is given by

$$\mu_D = A_1 \left\{\frac{f_2}{f_1}\right\}^2 + B_1 \left\{\frac{f_2}{f_1}\right\} + C_1 \tag{4.4}$$

where μ_D is dynamic Poisson's ratio, f_1, f_2 are first and second longitudinal resonance frequencies (in Hz), and A_1, B_1, C_1 are constants based on the dimension of the cylindrical specimen. Dynamic modulus is given by

$$E_D = 2(1 + \mu_D)\rho\left\{\frac{2\pi f_1 R_0}{f_n'}\right\}^2 \tag{4.5}$$

where R_0 is the radius of the cylindrical specimen. The control natural frequency at which the modulus is estimated is given by

$$f_n' = A_2[\mu_D]^2 + B_2[\mu_D] + C_2 \tag{4.6}$$

where the constants A_2, B_2, and C_2 are based on the dimension of the cylindrical specimen (Subramaniam et al., 2000).

For comparison of the dynamic modulus of elasticity with that of the elastic modulus, the following relationships hold good (Neville, 1997):

$$E = 1.25\, E_D - 1 \tag{4.7}$$

4.6 DETERMINISTIC AND PROBABILISTIC APPROACHES

4.6.1 Deterministic Approach

An approach based on the premise that an explicit and unique solution is available is a deterministic approach. In this approach, uncertainties are not formally recognized or accounted for; hence, uncertainties are associated with a probability of either 0 or 1. Although safety factors indirectly account for the uncertainties, one is concerned only with the reliability associated with this value. In conventional analysis, however, one is not concerned with the reliability associated with this unique value.

4.6.2 Probabilistic Approach

A probabilistic approach is based on the concept that several or varied outcomes of a situation are possible; hence, there is no unique solution for a given problem. Uncertainties are formally recognized in this approach. Description of the physical system includes randomness of the data and other uncertainties. This approach aims to determine only the probability of the outcome of any event that may occur. It is expressed in percent, indicating the degree of confidence in the estimated values. Probabilistic modeling aims to study a range of outcomes from given input data. Accordingly, the description of a physical situation or system includes randomness of data and other uncertainties. The selected data for a deterministic approach will, in general, not be sufficient for a probabilistic study of the same problem.

Reliability analysis cannot be accurate because it is practically impossible to identify all uncertainties; hence, engineering judgment is applied. Further, the method of modeling and analyzing them is not easy. Therefore, many assumptions are made during the analysis, which is the primary reason for inaccurate results. Most importantly, analytical formulation of the limit state surface and subsequent integration of the probability density function within the domain of interest is a highly complex task that declines the accuracy of the reliability estimates (Naess and Moan, 2013).

4.7 FORMULATION OF A RELIABILITY PROBLEM

Formulation of a reliability problem can be broadly divided into two groups: (i) time-invariant problems and (ii) time-variant problems. In both cases, limit state function, which could be based on the serviceability requirements or ultimate strength criteria, is defined. The reliability problem seeks to find the probability of the limit state of failure or violation of limit state conditions (Papoulis and Pillai, 1991).

4.7.1 Time-Invariant Problem

Let $g(x)$ denote a limit state function, where x denotes a set of random variables $\{x_1,x_2,x_3,\ldots\}$. Failure event is given by

$$g(x) \le 0 \tag{4.8}$$

Probability of failure is then defined as

$$P_f = P[g(x) \le x] = \int f(x)dx \tag{4.9}$$

where $f(x)$ is the joint probability density function of $g(x) = 0$ and the integral is performed over the domain D, where $g(x) < 0$. Reliability is given by

$$R = 1 - P_f \tag{4.10}$$

The plot is shown in Figure 4.1.

4.7.2 Time-Variant Problem

A reliability problem is said to be a time-variant problem if the limit state function is also a function of time. Therefore, a typical limit state function is $g[x,y(t)]$, where x is the set of random variables $\{x_1,x_2,\ldots,x_n\}$ and $y(t)$ denotes a vector of stochastic process. Failure event of such time-variant problems may constitute the out-crossing of the vector $y(t)$ through the limit state surface $g(x,y) = 0$. A typical plot is shown in Figure 4.2.

Probability density function is given by

$$P_f = \int_{0 \le t \le T} P[\{\min g(x,y(t) \le | x\}]f(x)dx$$

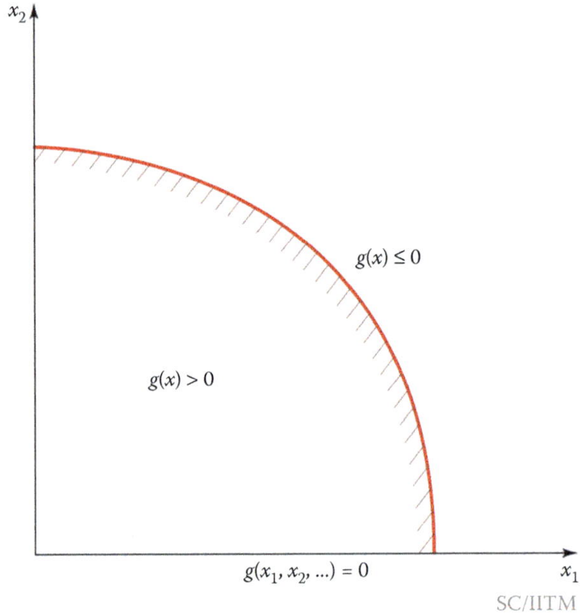

FIGURE 4.1 Failure domain.

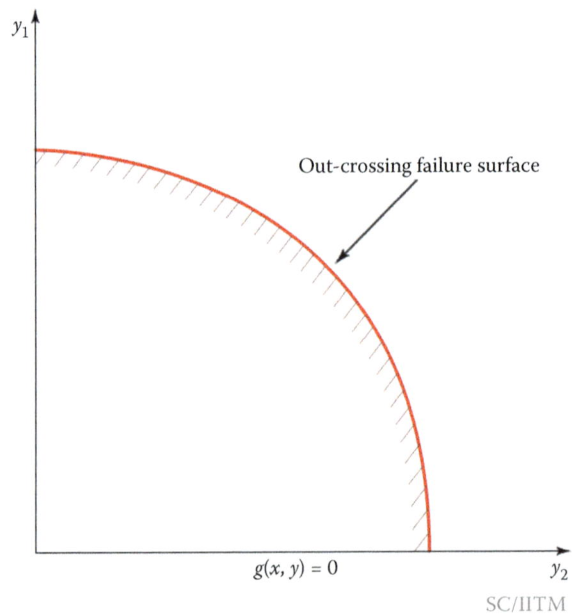

FIGURE 4.2 Domain indicating out-crossing failure.

where T denotes the life time of the structure. Please note that reliability is applicable to structures only to assure the satisfactory condition for a specified period (which in this case is T); $f(x)$ denotes the probability density function of (x). Conditional probability is obtained by stochastic analysis for random loading $y(t)$. Exact solution of the out-crossing analysis is very difficult for two reasons: (i) evaluation of conditional probability and (ii) determination of joint probability density function. Hence, the exact solution of the above equation is not possible; various levels of approximations are used in reliability analysis problems. Depending upon how the joint probability density functions are calculated (whether analytical or numerical), reliability methods are classified in the literature.

4.8 RISK AND RELIABILITY

In engineering practice, one routinely encounters situations that involve an event which has a particular probability of occurrence and, if it occurs, has a specific set of consequences (Schuêller and Choi, 1977). With experience and engineering judgment, one should be able to assign a suitable probability to the occurrence of an event, and some quantified magnitude or cost to the adversity associated with its occurrence (Chandrasekaran and Saha, 2011; Chandrasekaran, 2013a,b,c; Cornell and Vanmarche, 1975). This combination of uncertain event and adverse consequence is the determinant of risk. In engineering practice, to assess risk, two major inputs are required: (i) probability of occurrence of the event and (ii) consequences of the event happening. The product of these two is the risk, which is given by

$$\text{Risk} = (\text{probability of failure}) * (\text{consequence})$$

When more than one event is involved, then it can be the sum of the products of all such events, as given below:

$$\text{Risk} = \sum_i P_i c_i$$

Another useful measure here is reliability, which is defined as (1 – probability of failure), and expresses probability of *safety*. It is expressed as a reliability index (β), which is useful in describing the boundary between safe and unsafe conditions. *Therefore, risk is an extension of reliability to address the consequences of failure; reliability terminates at the stage of assessing the probability of failure, while risk continues to address beyond this point.* It is important to know that risk conveys the financial component of unsafeness, which is required if one has to justify the cost investment toward risk mitigation or reduction; the latter is a common practice in oil and gas industries. Hence, risk covers a broad spectrum of adverse effects on the following: (i) society, (ii) financial status, (iii) human life, and (iv) processing plant as a whole.

4.8.1 Acceptable Risk

Largely in offshore industry, risk cannot be reduced to *zero*. On the other hand, that does not mean that all offshore installations are unsafe. There is a well-defined level of acceptable risk beyond which installations are declared unsafe. Risk acceptance criteria are predefined by international authorities and implemented in cooperation with the local legislative authorities. Regulatory agencies include: (i) Environmental Protection Agency (EPA), (ii) Nuclear Regulatory Commission (NRC), and (iii) Federal Energy Regulatory Commission (FERC). The procedures these regulatory agencies use to separate reasonable from unreasonable risks vary from highly analytical to qualitatively procedural.

4.9 LEVELS OF RELIABILITY

Methods for the reliability analysis of structures are classified based on the level of approximations made (Chandrasekaran et al., 2011). Level I methods are appropriate levels of reliability, and are equipped with a number of partial safety factors to account for uncertainties in loads and material properties (Røren and Furnes, 1976, 1977). Probability aspects of the problem are taken into account in level I. Suitable characteristic values of the random variables are introduced into the analysis with the main objective of minimizing the deviation of the design values from the target value. Idealizations are made in level II methods; basic variables are described only with first-order statistics. This also leads to the linearization of the limit state curves and the reduction of failure modes as well. Level III methods can be characterized as being exact probabilistic methods of analysis. They are based on the full statistical distribution of all relevant variables. Such methods are not yet applicable to marine structures due to lack of information on the important variables. This deals with the integration of a multidimensional probability density function of random variables that is extended over the safety domain. Reliability is expressed in terms of suitable safety indices. A level IV reliability method also addresses the engineering economics of strategic importance. It includes the cost–benefit analysis, rehabilitation, consequences of failure (in a larger picture), and return on investment. A few examples where level IV reliability is applicable include marine structures, nuclear power plants, coastal protection structures, among them. Although level I methods are still most often used in structural design, level II methods are useful in evaluating the safety of a structure. They can also be used as a tool to determine the rational set of safety factors. There are different levels of reliability analysis, which can be used in any design methodology depending on the importance of the structure (Pavia et al., 1977).

The term "level" is characterized by the extent of information about the problem that is used and provided. The methods of safety analysis proposed for the attainment of a given limit state can be grouped under four basic "levels" (namely, levels IV, III, II, and I) depending upon the degree of sophistication applied to the treatment of various problems. In *level I* methods, the probabilistic aspect of the problem is taken into account by introducing into the safety analysis suitable "characteristic values" of the random variables, conceived as fraction of a predefined order of the statistical distributions concerned. These characteristic values are associated with partial safety factors that should be deduced from probabilistic considerations to ensure appropriate levels of reliability in the design. In this method, reliability of the design is deviated from the target value with an objective to minimize the error. The load and resistance factor design (LRFD) method comes under this category. Reliability methods, which employ two values of each uncertain parameter (i.e., mean and variance), supplemented with a measure of the correlation between parameters, are classified as *level II* methods. *Level III* methods encompass complete analysis of the problem and involve integration of the multidimensional joint probability density functions of the random variables, which are extended over the safety domain. Reliability is expressed in terms of suitable safety indices, viz., reliability index β and failure probabilities. *Level IV* methods are appropriate for structures that are of major economic importance. This involves the principles of engineering economic analysis under uncertainty and considers costs and benefits of construction, maintenance, repair, consequences of failure, interest on capital, etc. Sensitive projects such as nuclear power projects, transmission towers, highway bridges, and offshore and marine structures are suitable objects of level IV design.

4.9.1 Space of Variables

For analysis, we need to define the state of variables of the problem. The state of variables involves the basic load and resistance parameters used to formulate the performance function. For "n" state variables, the limit state function is a function of "n" parameters. If all load effects are represented by the variable Q and total resistance (or capacity) by R, then the space of state variables is a two-dimensional space as shown in Figure 4.3. Within this space, we can separate the "safe domain"

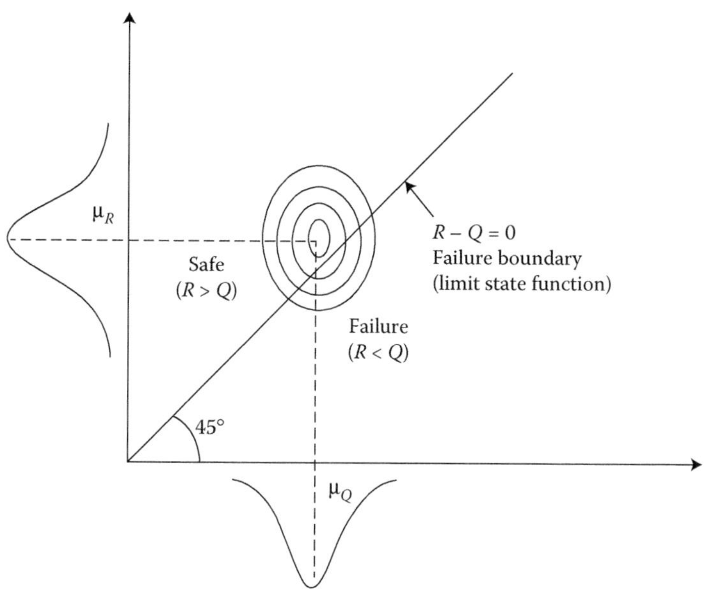

FIGURE 4.3 Safe domain and failure domain.

from the "failure domain"; the boundary between the two domains is described by the limit state function $g(R,Q) = 0$. Since both R and Q are random variables, we can define a joint density function $f_{RQ}(r,q)$. The probability of failure is calculated by integrating the joint density function over the failure domain [i.e., the region in which $g(R,Q) < 0$]. As noted earlier, this probability is often very difficult to evaluate and hence reliability index is used to quantify the structural reliability.

4.10 ADVANTAGES OF RELIABILITY METHODS

The major advantages of reliability methods are the following: (i) they offer a realistic procession of uncertainties; (ii) they offer a rational method to estimate safety, which is otherwise highly subjective; (iii) they offer decision-making support for noneconomic and better balanced design; (iv) they help identify zero failure nodes through the analysis; and (v) they provide measure by application of regulations, which would otherwise become irrational.

4.10.1 Difficulties in Reliability Models

Implementing reliability models for safety assurance of marine structures will have a few difficulties. The elementary variables that qualify uncertainties are material strength, geometry of the structural system (e.g., stiffness degradation due to ageing effects), and probabilities of load exceedance of loads within the service life. The first level of difficulty arises from the fact that variables belong to a type of uncertainty, which is different from one used to qualify for component-level variables (e.g., location behavior, environmental effects). The second level of difficulty is caused by the failure modes that are necessary to quantify the serviceability requirements (e.g., limit stress and limit displacement). The third level of difficulty arises from the separation of safety and failure domains. Classifying these domains, in terms of probability density function required for estimating the reliability index, is very complex. The fourth level of difficulty arises from the nature of random variables used to quantify the uncertainties of different events in the probabilistic model (e.g., external, internal, and their interdependency). Lastly, the major difficulties arise from the static or dynamic content of the loads (e.g., the influence of time on the overall process).

4.10.2 Steps in Reliability Studies

Reliability studies are carried out in different steps: (i) element level, (ii) component level, (iii) system level, and (iv) field investigation. While the first step is solved using stochastic modeling, further steps are usually carried out using probabilistic models to address either the component-level failure or the system-level failure as the case may be. Although field investigations do not demand any specific type of tool to conduct the investigations, data processing to qualify the ensemble of variables and enable a successful probabilistic analysis is a challenge. It is more difficult in the domain of application to marine structures for a simple reason that data based on these grounds are very weak.

4.11 RELIABILITY FRAMEWORK IN MARINE STRUCTURES

Structural reliability, as applied to marine structures, will have bifold meaning: (i) general sense and (ii) mathematical sense. In the general sense, reliability of the marine system is to ensure its ability to fulfill the design requirements for a specified period of time under specific conditions (Naess and Moan, 2013; Schuêller and Choi, 1977). In the mathematical sense, it is the probability that the marine system will not attain the specified limit state (serviceability or ultimate) during the specified reference period. Probability of safety is given by $(1 - P_f)$, which is also equal to Prob $(R > S)$, where R represents the resistance and S the load effects. Based on how R and S are related to time, the probability of failure will also have a reference period. Therefore, the basic problem of reliability is to *examine the probability of failure*. If the load effects (S) and the resistance (R) are expressed by the respective probability density functions $f_s(S)$ and f_R, then the probability of failure is given by

$$P_f = \text{Prob}(R \le S) \tag{4.11}$$

$$= \int_0^\infty F_R(s) f_s(s) ds = F_m(0) \tag{4.12}$$

where M is called the margin of safety, which is given by

$$M = R - S \tag{4.13}$$

If PDF $\{f_m(m)\}$ or CDF $\{F_m(m)\}$ are known, then the probability of failure can be computed either analytically or numerically.

4.11.1 When Both *R* and *S* Are Normally Distributed

When resistance and load effects are normally distributed, then the following relationships hold good:

$$P_f = \varnothing(-\beta) \tag{4.14}$$

where β is the reliability index and is given by

$$\beta = \frac{\mu_R - \mu_S}{\sqrt{\sigma_R^2 - \sigma_S^2}} \tag{4.15}$$

4.11.2 When Both *R* and *S* Are Log-Normally Distributed

When both resistance and load effects are log-normally distributed, then the reliability index is given by

$$\beta = \beta_{LN} = \frac{\ln\left[\mu_R/\mu_S\sqrt{(1+\gamma_S^2)/(1+\gamma_R^2)}\right]}{\sqrt{\ln(1+\gamma_R^2)(1+\gamma_S^2)}} \tag{4.16}$$

$$\beta_{LN} \approx \frac{(\mu_R/\mu_S)}{\sqrt{\gamma_R^2+\gamma_S^2}} \tag{4.17}$$

where γ_R and γ_S are resistance factor and load factors, respectively. In the above equation, reliability index is referred to a period. For example, if probability of failure is annual, then load effects should also refer to the annual maximum value and resistance should refer to the annual minimum value.

4.11.3 Implicit Failure Probability in Design

Let the random load effect (S) and resistance (R) be defined by

$$\mu_S = B_S S_C \quad \{\text{for } B_S \le 1.0 \quad \text{and} \quad \gamma_S = 0.15 \text{ to } 0.30\} \tag{4.18}$$

$$\mu_R = B_R R_C \quad \{\text{for } B_R > 1.0 \quad \text{and} \quad \gamma_R > 0.10\} \tag{4.19}$$

where B_s reflects the ratio of the mean load, which will refer to the annual maximum value instead of mean load if the probability of failure is annual, and S_c refers to characteristic load effect, which is a typical value over a period of 100 years. For (R,S) to be log-normal, reliability index is given by

$$\beta_{LN} = \frac{\ln(\mu_R/\mu_S)}{\sqrt{\gamma_R^2+\gamma_S^2}} = \frac{\ln(B_R V_R V/B_S)}{\sqrt{\gamma_R^2+\gamma_S^2}} \tag{4.20}$$

4.12 ULTIMATE LIMIT STATE AND RELIABILITY APPROACH

Ultimate limit states are governed by the design because they are generally based on extreme load effects. Ultimate strength formulations, which are supported by extensive test results, are traditionally based on the strength of material. Therefore, design criteria and reliability analysis will be based on limit state function.

4.12.1 Single *R* and *S*

For a single resistance and corresponding load effect (R and S), limit state function $g(x)$ is given by

$$\frac{R_c}{\gamma_R} \geq \gamma_{s1} S_{1c} + \gamma_{s2} S_{2c} \tag{4.21}$$

where the subscript "*c*" refers to the characteristic value, *R* is resistance, *S* is load effect, γ_R is the resistance factor, and γ_{S1}, γ_{S2} are load factors. Characteristic strength of the material refers to 5% fraction of the material strength, while characteristic value of the loads refers to the magnitude of the environmental load with the annual probability of exceedance of 10^{-2}. Design criteria are now given by $g(R_d, S_{1d}, S_{2d}) > 0$, where the following conditions also apply:

$$R_d = \frac{R_c}{\gamma_R} \tag{4.22}$$

$$S_{1d} = \gamma_{S1} S_{1C} \tag{4.23}$$

$$S_{2d} = \gamma_{2c} S_{2c} \tag{4.24}$$

Expression for the failure event is given by

$$g(R, S_1, S_2) \leq 0 \tag{4.25}$$

4.12.2 Multiple *R* and *S*

A structural member can be subjected to a different load combination of axial load and bending moment; then the analysis is required to be carried out for multiple resistance and load effects. In such cases, the failure criteria can be formulated as follows:

$$g(R_1, R_2, R_3, S_{1j}, S_{2j}) = 1 - \left[\frac{S_{ij}}{R_1} + \frac{S_{2j}}{(1 - (S_{1j}/R_2))R_3} \right] \tag{4.26}$$

It can also be rewritten as

$$g(R_1, R_2, R_3, S_{1j}, S_{2j}) = 1 - \left[\frac{X_1}{X_1} + \frac{X_3}{(1 - (X_1/R_4))} \right] \tag{4.27}$$

where S_{1j}, S_{2j}, etc. are the load effects for different combinations and *R* is the resistance while the count *j* stands for the load type. Equation 4.27 is based on the Perry–Robertson approach in which R_1, R_2 are the axial forces and moment capacity and R_3 is the Euler load. In professional practice, characteristic values of loads and resistance are used while they are considered as random variables in the reliability analysis. Partial safety factor γ_{S1} is computed using the reliability theory so that the design equation given above corresponds to the target failure probability (Haldar and Mahadevan, 2000).

4.13 SHORT-TERM RELIABILITY OF SINGLE LOAD EFFECTS

If the resistance (R) is constant over time and the load effect (S) is arising from a single source, then the characteristic value of the load effect can be obtained from the distribution of individual maxima of the largest value in that given period of time. In that case, the reliability problem can be derived based on the extreme value statistics to characterize S_{max}. Failure probability in the short time period $(0,T)$ is given by

$$P_f(T) = \text{Prob}(g, \max(Q(T)) \leq 0 \tag{4.28}$$

$$P_f(T) = \text{Prob}\,(g(R, S_{max}(t)) \leq 0 \tag{4.29}$$

where $g(.)$ is the limit state function, R is the resistance, and S is the load effect resulting from the process $Q(T)$.

4.13.1 Up-Crossing Approach

Alternatively, failure function can be expressed by up-crossing approach, in which the up-crossing rate is established. In case of f time-dependent reliability problems, the main interest lies in the time (t_f) to that of the first failure. For a simple problem, the following relationship holds good:

$$M(t) = g(R, S(t)) = R - S(t) \tag{4.30}$$

where t_f is the first time when $M(t) = 0$. Time to first failure (t_f) is the time of the first excursion of $M(t)$ from a positive value to a negative value, assuming $M(t)$ is a continuous process. Probability of failure in the period $[0,T]$ is equivalent to the probability of $t_f < T$, which is given by

$$P_f(T) = 1 - \text{Prob}(t_f > T) \tag{4.31}$$

$$P_f(T) = 1 - \text{Prob}(N(T) = 0 | M(0) > 0)\,\text{Prob}(M(0) > 0) \tag{4.32}$$

where $N(T)$ is the number of up-crossings within the interval $(0,t)$. This will be equal to the number of crossings from the safe domain to the failure domain. $M(0) > 0$ signifies that the safety margin $M(t) = R - S(t)$ is in the safe domain at zero time. In general, calculating $P_f(T)$ is a complex task. An approximate solution can be achieved by assuming $N(t)$ as a Poisson process. In that case, up-crossings of level R by $S(t)$ are independent of the mean rate of the following:

$$\gamma_S^+(R) = \gamma_M^-(0)\,\text{per unit time} \tag{4.33}$$

Hence, the following is valid:

$$\text{Prob}\,(N(T)=0)=\frac{(\gamma_S^+(R)T)^0}{0!}e^{-\gamma_S^+(R)T}=e^{-\gamma_S^+(R)T} \tag{4.34}$$

Further, the following statement declares the probability of no failure at $t = 0$:

$$\text{Prob}(M(0)>\,)=1-P_f(0) \tag{4.35}$$

Hence, $P_f(0) = 0$.
Then, $P_f(t)$ is given by

$$P_f(t)\approx 1-e^{-\gamma_S^+(R)T} \tag{4.36}$$

$$\cong -\gamma_S^+(R)T=-\gamma_M^-(0)T \tag{4.37}$$

As a special case, if $S(t)$ is a Gaussian process, then the following relationship holds good:

$$\gamma_S^+(R)=\gamma_0^+e^{-[(R-\mu_S)^2/2\sigma_S^2]} \tag{4.38}$$

where $\gamma_0^+=\gamma_S^+(0)$.

It is also important to note that various uncertainties other than the fundamental variability in loading also affect the up-crossing rate. For the given values of random variables X ($= x_1,x_2,\ldots,x_n$) that represent those uncertainties, conditional probability of failure is determined by the down-crossing of zero by $M\,(t;X)$ for ≥0. It is given by

$$P_f(x)=\text{Prob}\,(\min_{0\le t\le T}M(t\,|\,x)\le 0) \tag{4.39}$$

$$=1-e^{-[\gamma_m^-(0;x)T]} \tag{4.40}$$

where the exponential power indicates zero down-crossing rate, which depends on the parameter, $\{x\}$. Total failure probability, considering the uncertainties in $\{X\}$, can be computed by the unconditional probability as given below:

$$P_f=\int_x P_f(x)f_X(x)\,dx \tag{4.41}$$

The integral represents the expected value of $P_f(X)$. Hence, it can be calculated as an average of $P_f(x_i)$ when the outcomes of x_i are generated by Monte Carlo simulation. Hence, the probability of failure is rewritten as

$$P_f\cong\frac{1}{N}\sum_{i=1}^{N}P_f(x_i) \tag{4.42}$$

4.14 LONG-TERM RELIABILITY OF SINGLE LOAD EFFECT

For a nonstationary process $M(t;x)$, which could be for a long-term period, the failure probability is given by

$$P_f(x) = 1 - e^{-\int_0^T \gamma_M^-(0;t,x)\,dt} \tag{4.43}$$

where $\gamma_M^-(0;t,x)$ is the mean down-crossing rate, which depends on the sea state that changes with time. For a given sea state, let W be the captured values. Then, the following relationship holds good:

$$W = f(H_s T_p U_s U_w, \alpha) \tag{4.44}$$

where H_s is the significant wave height, T_p is the spectral peak period, U_c is current velocity, U_w is wind speed, and α is the wave, wind, and current directions, respectively (Yang and Freudenthal, 1977). The prerequisite of long-term probability of failure is to impose an ergodicity assumption on the environmental process as below:

$$W = W(t) \tag{4.45}$$

Then, the probability of failure is given by

$$P_f(x) = 1 - \exp\left\{-T\int \gamma_m^-(0;w,x) f_w(w)\,dw\right\} \tag{4.46}$$

where $\gamma_m^-(0;w,x)$ denotes mean zero down-crossing rate for M for the sea state $(W = w)$; the following relationship holds good:

$$f_W(w) = \text{PDF of } w \tag{4.47}$$

The total long-term failure probability, accounting for both the environmental variability and parameter uncertainty is given by

$$P_f = \int P_f(x) f_X(x)\,dx \tag{4.48}$$

EXERCISE PROBLEMS

1. Out of 850 manufactured parts of a machinery component, 50 do not confirm to the requisite standards; 2 parts are taken at random without replacement from the batch to assess the quality check. Find the cumulative mass function.

   Let the random variable be X, which is equal to the nonconfirming part in the sample.

Probability mass function of X

$P(X = 0) = (800/850) \times (799/849) = 0.884$
$P(X = 1) = (800/850) \times (50/849) \times 2 = 0.110$
$P(X = 2) = (50/850) \times (49/849) = 0.003$

Cumulative mass function

$F(0) = P(X <= 0) = 0.884$
$F(1) = P(X <= 1) = 0.884 + 0.11 = 0.994$
$F(2) = P(X <= 2) = 0.994 + 0.003 = 1$

2. A, B, and C are three people waiting for placements in a company. Only one job vacancy is available. Data say that A has 0.5 chance of that of B; B has 0.667 of that of C. What is the probability that A, B, or C shall get the job?
   $P(A) + P(B) + P(C) = 1$; They are mutually exclusive.

Given that

$P[A] = (1/2) \times P[B]$
$P[B] = (2/3) \times P[C]$
$P[C] = (3/2) \times P[B]$

Therefore, $(1/2)\ P[B] + P[B]\ + (3/2)P[B] = 1$;

$2P[B] = P[B] = 1$
$P[B] = (1/3) = 33\%$
$P[C] = 51\%$
$P[A] = 17\%$

3. Out of three accelerometers available in the lab, find the probability that at least one accelerometer shall be working after 1 year?
   Assigning the probabilities as G for good and B for bad, simply by possible combinations, one can assess the required probability. To start with, all the three accelerometers were new and shall be working well. Therefore, G is assigned for all. Pick up the probability that at least one of them is in working condition.

Accelerometer 1	Accelerometer 2	Accelerometer 3
G	G	G
G	G	B
G	B	G
B	G	G
G	B	B
B	G	B
B	B	G
B	B	B

P (1 accelerometer after 1 year) = (3/8)

4. In the test conducted on concrete mix design, it is concluded that 30% of coarse aggregate, 25% of fine aggregate, 25% of cement, and 20% of water

generally control the overall strength. These data were derived based on the experiments conducted in the lab. However, it is understood that not all the samples tested were reliable and therefore one cannot rely on the statement made for any further assessment. It is now necessary to examine the adequacy of the test results; a sample having the following data is to be assessed. Course A = 27%; FA = 10%; cement = 30%; water = 50%. What is the reliability of information from the sampling on one of the random samples?

Solution

We shall use total probability theory to solve this problem:

$$p[A \mid B] = p[A]p[B \mid A]$$

$$\begin{aligned} P[\text{conc} \mid B] &= p[\text{conc}] \times p[B \mid A] + p[\text{conc}] \times p[B \mid CA] + p[\text{conc}] \\ &\quad \times p[\text{conc} \mid \text{cement}] + p[\text{conc}] \times p[\text{conc} \mid \text{water}] \\ &= 0.3 \times 0.27 + 0.25 \times 0.1 + 0.25 \times 0.3 + 0.2 \times 0.5 \\ &= 0.281 \end{aligned}$$

5. A sample of study was conducted to estimate the *in situ* strength of offshore structures. The sampling was done only on concrete platforms. Destructive testing was followed and core cuttings were made. Test standards suggest that the concrete core should be rejected if at least one out of four is not satisfactory. Estimate the probability that the core cut from the platform is accepted if 100 samples are available and gives poor results?

Let $E(i)$ be the probability of finding that the sample is accepted. For the four core being cut

$$\begin{aligned} P[\text{acceptance}] &= \frac{95}{100} \times \frac{94}{99} \times \frac{93}{98} \times \frac{92}{97} \\ &= 0.812 \end{aligned}$$

For 50 samples

$$\begin{aligned} P[\text{acceptance}] &= \frac{45}{50} \times \frac{44}{49} \times \frac{43}{48} \times \frac{42}{47} \\ &= 0.65 \end{aligned}$$

For 25 samples

$$\begin{aligned} P[\text{acceptance}] &= \frac{20}{25} \times \frac{19}{24} \times \frac{18}{23} \times \frac{17}{22} \\ &= 0.38 \end{aligned}$$

For 200 samples

$P[200] = 0.9$

4.15 RELIABILITY METHODS

4.15.1 First-Order Second-Moment Method

First-order Taylor series approximation of the limit state function is used in this method (Rackwitz, 1977). Only second moments of the random variables are used to estimate the probability of failure. Limit state function is defined as

$$M = R - S \tag{4.49}$$

where R and S are statistically independent and normally distributed. In that case, M is also normally distributed. Then, the following are valid:

$$\mu_m = \mu_R - \mu_S$$

$$\sigma_m = \sigma_R^2 + \sigma_S^2 \tag{4.50}$$

Probability of failure $P_f = P(M < 0)$

$$= P[(R - S) < 0]$$

If M is the normal variant, then

$$P_f = \phi\left(\frac{-\mu_m}{\sigma_m}\right) \tag{4.51}$$

$$\beta = \text{reliability index} = \frac{\mu_m}{\sigma_m} \tag{4.52}$$

ϕ is the case cumulative distribution function of standard normal variable

$$P_f = 1 - \phi\left(\frac{\mu_R - \mu_S}{\sigma_R^2 + \sigma_S^2}\right) \tag{4.53}$$

If R and S are log normal, then the probability of failure is given by

$$P_f = 1 - \varnothing\left[\frac{\ln\left[(\mu_R/\mu_S)\sqrt{(1 + V_S^2)/(1 + V_R^2)}\right]}{\sqrt{\ln(1 + V_R^2)(1 + V_S^2)}}\right] \tag{4.54}$$

FOSM has a few advantages: (i) it is easy to use and (ii) it does not require the knowledge of distribution of the random variables. However, disadvantages include the following: (i) results are inaccurate if the tail of the distribution function cannot be approximated by normal distribution and (ii) the value of the reliability index depends on the specific form of the limit state function, which is a classical *invariance problem.*

4.15.2 Advanced FOSM

It is necessary that all mechanically equivalent performance functions must produce the same safety indices, irrespective of evaluation of the limit state function. This is required for a reliability estimate of multiperformance functions, which is very common in marine structures due to the integration of various functional and structural components. The reliability index provides the exact probability of failure only when all the random variables used to describe the performance of the system are statistically independent and normally distributed. This implies a prerequisite that the performance function should be a linear combination of random variables. The Hasofer–Lind method is an advanced improvement on FOSM, which eliminates this complexity. The method centers around computation of *design point.* Design point is the point of minimum distance of the performance function from the origin. This minimum distance is shown to be the reliability index for the case of random variables that are statistically independent and normally distributed (and also performance function remains as a linear combination of random variables). The method actually transforms the random variable into a reduced form as given by

$$x_i' = \frac{x_i - x_i}{\sigma_{xi}} \quad \{\text{for } i = 1,2,\ldots,n\} \tag{4.55}$$

Thus, the reduced variables shall have a zero mean and unit standard deviation. Now, the performance function $(G(x) = 0)$ will be transformed to $(G(x') = 0)$. The minimum distance of the performance function from the origin will give the reliability index:

$$\beta_{HL} = \sqrt{x_d' \, x_d'^T} \tag{4.56}$$

where x_d' is the minimum distance of the design point from the origin, which is also referred as *check point.*

Case 1: When the limit state function is linear

Let us consider $M = R - S = 0$.

The reduced values are

$$R' = \frac{R - \mu_R}{\sigma_R}$$

$$S' \frac{S - \mu_S}{\sigma_S} \tag{4.57}$$

Substituting Equation 4.57 in Equation 4.56, we get

$$M = (\sigma_R R' + \mu_R) - (\sigma_S S' + \mu_S) \tag{4.58}$$

Figure 4.4 explains the limit state function.

From the figure, it is seen that as the limit state function is near to the origin, the failure region is larger and hence the probability of failure is higher. The minimum distance of the limit state (which is represented as straight line, as the performance function is linear) from the origin will give the reliability index as

$$\beta_{HL} = \frac{\mu_R - \mu_S}{\sqrt{\sigma_R^2 + R\sigma_R^2}} \tag{4.59}$$

The above expression is a simulator to that of reliability index of normal variate.

Case 2: When the limit state is nonlinear

When the limit state function is nonlinear, computation of the minimum distance becomes an optimization problem, as given below:

$$\beta_{HL} = D = \sqrt{(X')^T X'} \tag{4.60}$$

The above function is to be minimized for the condition that $G(x') = 0$. Many random variables, represented by the vector $\{X = X_1, X_2, \ldots, X_n\}$ that originates from the safe state $G(X') < 0$, are indicated as failure. Hence, $G(X') > 0$ is denoted as the minimum distance from the origin to the *design point* on the limit state, in the reduced

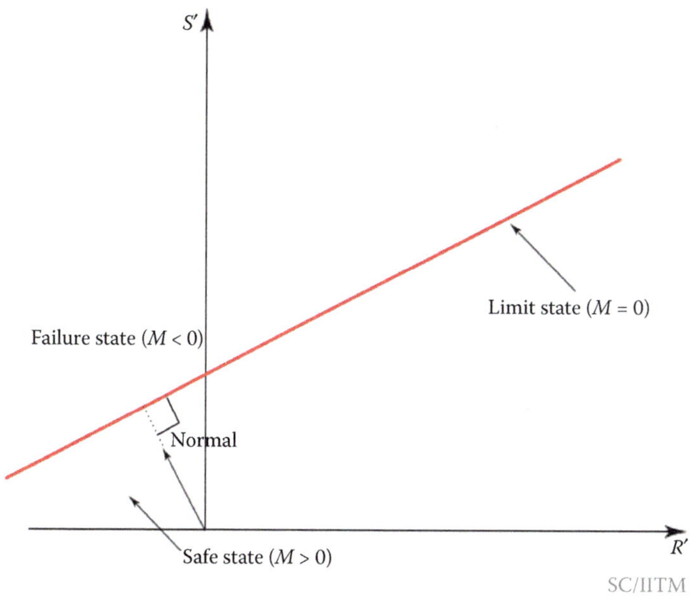

FIGURE 4.4 Limit and failure state.

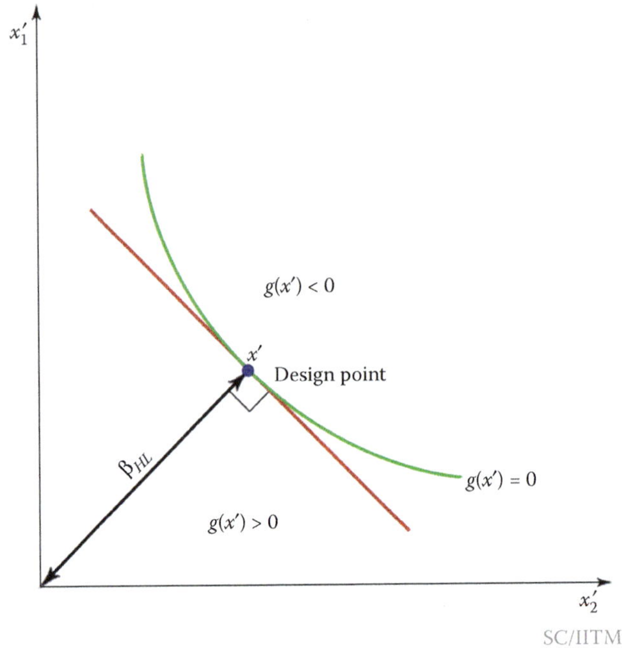

FIGURE 4.5 Reliability index and design point.

coordinate system. In this definition, the reliability index is invariant because, regardless of the form in which the limit state equation is written, its geometric shape and the distance from the origin remain constant, as shown in Figure 4.5.

From Figure 4.5, it is seen that x' is the most probable point or the design point. The Hasofer–Lind reliability index can be used to calculate the first-order approximation of the failure probability, as given below:

$$P_f = \Phi(-\beta_{HL}) \tag{4.61}$$

This is an integral of the standard normal density function along the ray joining the origin and x'. It is obvious to note that the nearer the x' is to the origin, the greater is the failure probability (Thoft-Christensen, 1977; Thoft-Christensen and Baker, 1982). Thus, the minimum distance point on the limit state surface is also called as the most probable failure point. Point of minimum distance x' represents the worst combination of the variables. The problem is now reduced to determine the coordinates of D, geometrically or analytically. Minimizing D (given in Equation 4.60) under the constraint that $G(X) = G'(X) = 0$, one can determine the distance using Lagrange multipliers as given below:

$$\beta_{HL} = -\frac{\sum_{i=1}^{n} x'_{di}\{\partial G/\partial x'_{di}\}}{\sqrt{\sum_{i=1}^{n}(\partial G/\partial x'_{di})^2}} \tag{4.62}$$

where $\partial G/\partial x'_{di}$ is the partial derivative, evaluated at the design point with coordinates $\{x'_{d1}, x'_{d2}, x'_{d3},\ldots, x'_{dn}\}$.

The design point is given by

$$x'_{d1} = -\alpha_{di}\beta_{HL}\{\text{for } i = 1,2,\ldots,n\} \tag{4.63}$$

$$\alpha_{di} = \frac{\{\partial G/\partial x'_{di}\}}{\sqrt{\sum_{i=1}^{n}(\partial G/\partial x'_{di})^2}} \tag{4.64}$$

Equation 4.64 gives the direction cosines along the coordinate axes, x'_i. Therefore, in the space of the original variable, design point can be determined using the following relationship:

$$x'_{d1} = \mu_{xi} - \alpha_{di}\sigma_{xi}\beta_{HL} \tag{4.65}$$

5 Fatigue and Fracture

5.1 INTRODUCTION

Fatigue design of marine structures is important when the dissipation of the long-term variation of local stresses is addressed. Long-term variations of local stress are generally caused by environmental loads, predominantly by waves. The main contribution to fatigue damage comes essentially from these load effects, which are in the order of 10%–20% of the effects caused by extreme loads during the service life of the structure. Fatigue failures are catastrophic; they cause significant damages without warning. Physical problems of fatigue consist of (i) initiation of crack, (ii) stable crack growth, and (iii) unstable crack growth until rupture. Once the crack is initiated, it tends to propagate in an orthogonal direction of tensile stresses developed.

Fatigue is a challenging failure mode with which to deal because of the following factors: (i) initiation process of fatigue is highly unpredictable, (ii) mapping of experimental studies done on lab scale to that of the field scale is difficult and sometimes erroneous, and (iii) modeling the environment process responsible for fatigue is not perfect. Fatigue failure is controlled by (i) addressing the probability of crack initiation in the design, material selection, and structural detailing stages, (ii) scheduling regular inspection during construction and operation, and (iii) following repair procedures as recommended by the design codes and classified societies. It is interesting to note that limit state conditions of design include cases of failure at some predefined extreme loads, which will include the fatigue life requirements. Commonly checked limit states for marine structures are (i) strength under extreme loading, (ii) fatigue life under slow crack growth under cyclic loading, (iii) fracture occurring under rapid growth under high amplitude loads, and (iv) limiting deflection, which is one of the important serviceability requirements. It is a common practice to present the results of strength, fatigue, and fracture checks as *unity check*. The unity check for strength is expressed as the ratio of actual or factored load to the acceptable load, whereas for fatigue, it is given by

$$U = \sqrt[m]{\text{Design life damage}}$$

where m is the negative slope of the S–N curve.

5.2 FATIGUE ASSESSMENT

One of the ways of fatigue assessment is based on the direct calculation of fatigue damage or expected fatigue life. Three important methods of assessment are (i) the simplified method, (ii) the spectral method, and (iii) the deterministic method. Alternatively, an indirect fatigue assessment can also be performed by limiting the predicted stress range to within the permissible stress range. Fatigue demand is stated in terms of the stress ranges that are produced by the cyclic loads, which are subsequently imposed on the structure. Such loads arise essentially from wave amplitudes, wave and wind directions, etc. As such loads are also dynamic in nature, the response amplitude of the structure will be dynamically excited under such loads. When considering fatigue-inducing stress ranges, one should also consider the stress concentration and its effect on modifying the active stress range. As a common practice, all analytical models may not adequately include the modification of stress ranges to address fatigue assessment; in such cases, analysis results are modified using a stress concentration factor (SCF). When a steel member is subjected to large fluctuating tensile stresses, small crack-like defects are initiated. They will grow in size and ultimately result in breaking (fracturing) of the member. For machinery design, these stress fluctuations are similar throughout their design life. Therefore, cyclic stresses are kept below the endurance limit stress to protect the machinery from failure. But for structural members, this method is not applicable due to the mixture of small and large amplitude stresses that arise from environmental loads. Common design approach is to use *S–N* curves, which are based on experimental investigations carried out on a variety of specimens. For checking purposes, design life (T, in years) is expressed as (N_T), the number of stress cycles expected within the design life of the overall structure. Calculated fatigue life, T_f (or N_f) is the computed life, in units of time (or number of cycles) for a particular structural member that corresponds to the appropriate *S–N* curve. Detection of fatigue failure is done from the occurrence of the visible crack and identifying through-thickness crack and loss of load-carrying capacity.

5.3 *S–N* APPROACH

Experiments conducted on steel specimens under fluctuating loads showed that the number of cycles to failure (N) is inversely proportional to the stress range (S) as given below:

$$N \propto \frac{1}{S^m} \tag{5.1}$$

$$N = AS^{-m} \tag{5.2}$$

where A and m are quantities derived experimentally. In marine structures, if control measures are undertaken, then it can be seen that there is a cut-off stress, below which fatigue failure will not occur, even under cyclic loading. This cut-off stress is found to be 2E08 cycles. For variable amplitude stress cycles, cut-off stress decreases with the growth of the crack. The slope of the curve changes from m to $(m + 2)$ beyond 1×10^7 cycles. This effect is significant if one is looking for a higher fatigue life. Hence, for preliminary analysis, the *S–N* curve can be taken as linear, which often simplifies the calculation. Over the range where the slope is constant, the *S–N* curves are plotted as a straight line on a log–log scale for a slope of $(-1/m)$. Since the *S–N* curves are experimentally plotted, local stress concentration effects caused by the shape of welds and of the specimen reduce the constant in Equation 5.2. Increases in thickness reduces the fatigue life and, hence, the constant, A, while fatigue life decreases with increases in freely corroding conditions. For marine structures, the *S–N* curves used in the design are given by

$$N = A_c S^{-m} \tag{5.3}$$

$$\log_{10} A_c = \mu - 2S \tag{5.4}$$

where μ and S are the mean and standard deviation of the parameter of $\log_{10} N$, respectively. Further, the reference parameter in the *S–N* curve is also expressed as

$$A = A_{ref}\left(\frac{t_{ref}}{t}\right)^{m/4} \tag{5.5}$$

where A_{ref} is the reference parameter in the *S–N* curve, t and t_{ref} are the thickness of the plate specimen and reference plate, respectively. *S–N* curves are traditionally determined by the constant amplitude testing with a large stress ratios (ratio of minimum to maximum stress is taken as 0.5), implying that the stress range considered in the experiments is in the crack opening mode. A few factors that make the existing *S–N* curve conservative are (i) relaxation of the residual stress, (ii) partial compression of the external stress, and (iii) crack closure effects. These factors would make the actual crack growth less than that implied by the stress ranges used in the existing *S–N* curves. While it is good to know that the approach is conservative, the bad news is that it deals with a constant amplitude, whereas the stresses are of variable amplitude due to the nature of the environmental loads that marine structures encounter.

5.4 MINER'S RULE

Miner's rule supplements the constant amplitude, which is not of serious concern when applied to the marine structures. This actually allows a number of different amplitude cycles, and the fatigue damage is based on this rule. Fatigue damage of joint under n cycles of constant amplitude is given by Equation 5.2 and the cycle is simply (n/N). If the joint is under a loading of variable amplitude, then the loading cycles can be divided into groups of approximately equal stress ranges. If there are G such groups with almost equal stress range in a given variable amplitude loading, then let S_g and n_g be the stress range and number of cycles in each group, respectively. Fatigue damage of each group is given by

$$D_g = \left(\frac{n_g}{N_g}\right) \tag{5.6}$$

$$N_g = AS_g^{-m} \tag{5.7}$$

Miner's rule states that failure under variable amplitude load will occur when the following condition is satisfied:

$$\sum_{g=1}^{G} D_g = 1 \tag{5.8}$$

This is often referred as *fatigue spectrum*.

5.5 FATIGUE LOADING AND FATIGUE ANALYSIS

Nominal stress range considered for fatigue assessment needs modification to account for the conditions that affect the local stress at that location (Kam and Dover, 1988, 1989). The ratio of the local to the nominal stress is defined as the SCF. Depending on specific situations, different SCFs may apply to different nominal stress components (Chaudary and Dover, 1985). It is most common to encounter SCF values larger than 1.0, signifying the amplification of the nominal stress. It is also important to note that when using this design of *S–N* curve in a fatigue assessment, applied reference stresses should correspond to the nominal stresses used in creating the curves. However, in an actual structure, it is rare that a match will be found with the geometry and loading of the tested specimens. Therefore, as a general rule, the following may be applied for the determination of the appropriate reference stresses required for a fatigue strength assessment:

- The effect of stress concentrations arising from weld profiles should be disregarded.
- Often, the *S–N* curve selected for the structural detail already reflects the effect of a stress concentration that arises from an abrupt geometric change. Therefore, there is no need to include this effect explicitly.
- If the stress field under current consideration is complex in comparison to a uniaxial field, the principal stress that is adjacent to the potential crack location should be used in the analysis.
- To include the assessment of structures under transverse loads, which is not the case for the *S–N* curve, reference stress should be taken as the nominal shear stress across the minimum weld throat area. This is necessary because in fillet welds, cracking could occur in the weld throat.

5.5.1 Fatigue Design

Local stresses for fatigue design need to be determined based on temporal and spatial variations. The stress concentration affecting various parts of the welded structure are as shown in Figure 5.1.

For welded structures, the stress range is the main parameter to represent the variation in time. The *S–N* approach is based on the fact that the tensile residual stresses are always present and that for all stress cycles, they drive the crack formation effectively. The spatial stress variation can also be accounted for by using the

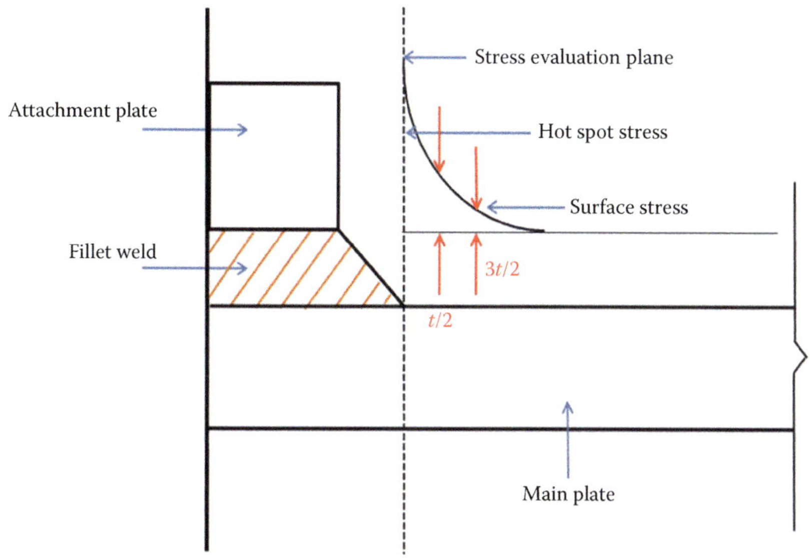

FIGURE 5.1 Sketch describing notch, hot spot, and surface stresses.

hot spot stress approach. Fatigue loading is essentially caused in marine structures by dynamic loads; the primary source is the wave loads. While global analysis of fatigue loading affects member forces, local analysis is carried out to determine the hot spot stresses. It is to be noted that the specified *S–N* curve used in the hot spot approach will not account for local geometric changes. Therefore, it is necessary to perform a detailed structural analysis to determine explicitly the stress concentrations due to such changes.

5.5.2 Stress Definitions

i. *Nominal stress,* S_{nom}: The stress at a cross section of the specimen or structural detail, which is measured away from the spot where a fatigue crack is initiated. Effects that arise from the geometry or weld profile are not considered in the nominal stress.
ii. *Hot spot stress,* S_{hot}: The stress at the surface is known as the hot spot stress. Changes in the stresses that arise from the weld profile are not included in the hot spot stress, but the overall effect of connection geometry on the nominal stress is represented.
iii. *Notch stress,* S_{notch}: This is the total stress at the weld toe. It includes the hot spot stress and the stress due to the presence of welds. Figure 5.2 shows the stress gradients near a weld, and Figure 5.3 shows the different types of hot spot stresses.

A *hot spot stress concentration factor* is defined as the ratio of the hot spot stress at a location to the nominal stress computed for that location. At sections where the "geometric" SCF is considered, a hot spot SCF may be obtained by directly

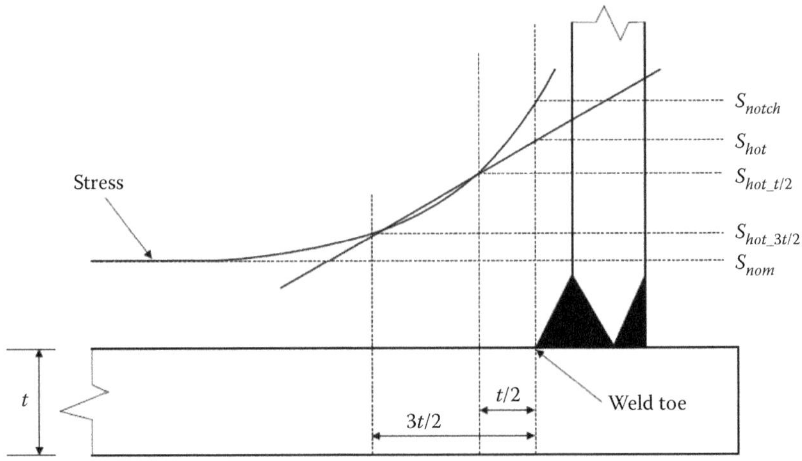

FIGURE 5.2 Stress gradients (actual and idealized) near a weld.

measuring the stress values using an appropriate physical model. As a common practice, designers use parametric equations to compute them for welded tubular joints in marine structures.

5.5.3 Hot Spot Stress Method

Stresses that are caused by axial forces and moments are called *primary hot spots*. They are caused by the structural details of the connections or the joint geometry. Damage in each group of connections is given by Equation 5.6. Based on the fatigue damage estimates made by different methods, the following statements are valid:

$$D_{RC} \leq D_{RFC} \leq D_{LCC} \leq\leq D_{PC} \tag{5.9}$$

$$D_{LCC} = D_{NB} \tag{5.10}$$

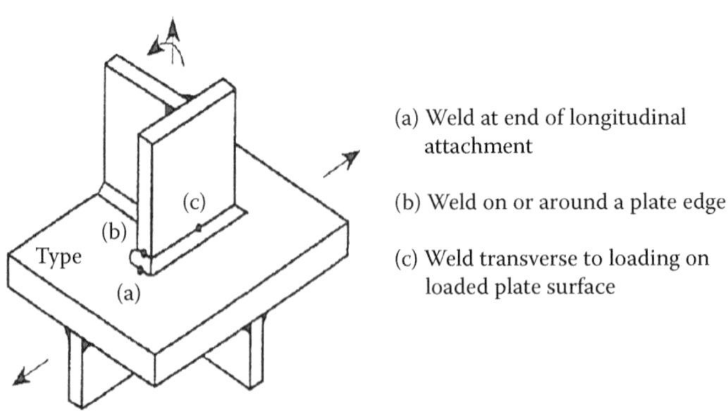

FIGURE 5.3 Types of hot spot stress.

where *RC* is range counting, *RFC* is rainflow counting, *LCC* is level crossing counting, *PC* is peak counting, and *NB* is narrow-band approximation. A hot spot stress approach is used when the toe of the weld is the location that is being assessed, where an attachment is also present. In such cases, the hot spot is considered to be the toe of the weld itself. Fatigue assessment of a tubular joint detail is performed on a hot spot stress basis, using *S–N* curves that apply for this purpose. Hot spot locations are the toes of the weld on both the chord and the brace sides of the weld; various other locations along the circumference may also be considered.

5.6 TIME-DOMAIN FATIGUE ANALYSIS

Time-domain fatigue analysis results in a time series of stresses. For a narrow-band Gaussian response, these cycles are well defined. Therefore, for more general stress–time histories, cycle counting methods are used to estimate the response in time series. One important factor that limits the use of this method is that it uses only the information provided by the series of peaks (local maxima) and valleys (local minima). There are different principles used to construct the effective stress ranges from the peaks and valleys reported in the time series; various counting methods, the procedures for which are well described in the literature, are used. These include (i) peak counting, (ii) range counting, (iii) level-crossing counting, and (iv) rainflow counting. Among all of them, rainflow counting is commonly used to estimate fatigue damage in marine structures.

5.6.1 Rainflow Counting

This method was first proposed by Matsuishi and Endo (1968). For a given stress time series of peaks and valleys, with the time axis vertical downwards, as shown in Figure 5.4, lines connecting the peaks and valleys form a series of *pagoda roofs*. Each rainflow starts at the beginning of the time series, and the point of origin is the peak and valley located inside. Rainflow, initiating at a peak (or a valley), falls down until it reaches the opposite peak, which is more positive (or a valley, which is more negative) than the peak (or the valley) from where it started. Rainflow also stops where it meets the rainflow from the roof above. Rainflow must terminate at the end of the time series. The horizontal length of each rainflow is counted as a half-cycle within that stress range.

5.6.2 Methodology

The methodology is based on standard principles (ASTM 2005; Wirsching and Ortiz, 2006).

- Reduce the time history to a sequence of peaks and troughs; peaks are considered as tensile.
- Each tensile peak is assumed as a source of water that falls down the pagoda.

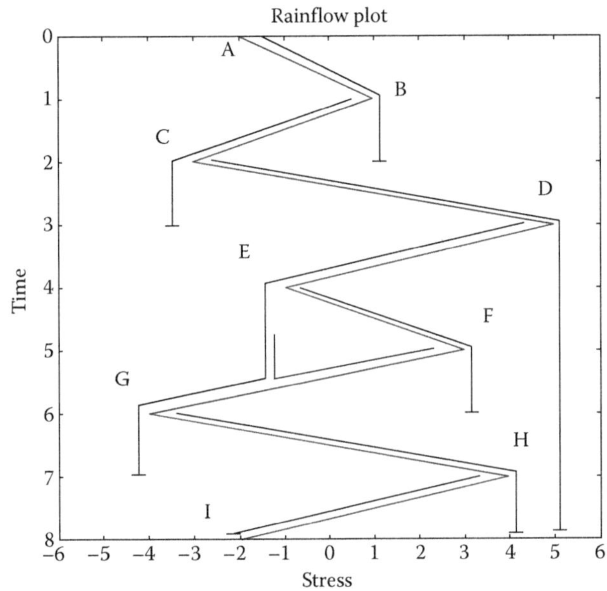

FIGURE 5.4 Rainflow counting method.

- Count the number of half-cycles by observing the terminations in the flow. Terminations in the flow occur due to either of the following reasons: (i) when it reaches the end of the series; (ii) when it emerges with a flow that started at an earlier peak or trough of a greater magnitude.
- Repeat the above steps for compressive troughs also.
- Assign a magnitude, which is equal to the stress difference between its start and termination, to each half-cycle.
- Pair up half-cycles of identical magnitudes but opposite in nature to count the number of complete cycles.

Referring to the figure, it is observed that the rainflow starts at point 1 (valley), 2 (peak), and 3 (valley) and falls down. The cycle ends at point 10, which is found to be a peak. There are nine half-cycles that can be extracted from the series. We can also observe that the rainflow that initiates from valley point 1 falls down to point 2 and subsequently ends at point 4, as the following valley has a smaller magnitude in comparison to the initiating point 1; hence, the half-cycle 1–2–4 is identified. A similar rule is applicable to the half-cycle 5–6. A second rainflow starts at point 2 and ends at point 3, as the successive peak at point 4 is larger in magnitude in comparison to point 2. Half-cycles 4–5–7 and 8–9 are also extracted in a similar manner. Half-cycle 7–8–9 is formed as the time series ends at point 10. Half-cycles 3–2′, 6–5′, and 9–8′ are determined as the rainflow starting at points 3, 6, 9 peaks meeting the rainflow from roofs above. When all of the half-cycles are exhausted, the horizontal length of each cycle is used as an effective stress range to calculate the fatigue damage based on the linear damage accumulation law.

Rainflow Cycles by Path

Path	Cycles	Stress Range
A–B	0.5	3
B–C	0.5	4
C–D	0.5	8
D–G	0.5	9
E–F	1.0	4
G–H	0.5	8
H–I	0.5	6

Rainflow, Total Cycles

Stress Range	Total Cycles	Path
10	0	–
9	0.5	D–G
8	1.0	C–D, G–H
7	0	–
6	0.5	H–I
5	0	–
4	1.5	B–C, E–F
3	0.5	A–B
2	0	–
1	0	–

5.7 DETERMINISTIC FATIGUE ANALYSIS

Deterministic fatigue analysis applies Miner's rule. Loading of the structure is represented by loading cases $\{g = 1 \text{ to } G\}$, each with a defined number of cycles (n_g) and time (T). Structure is analyzed to determine the stress (S_g) for each group and hence the total damage (D_T) in time (T) is estimated. If the value of time chosen for the analysis is 1 year, then the fatigue life of the structure is estimated as ($1/D_{year}$). When estimated fatigue damage is higher, a majority of the damage occurs on the low-cycle end of the S–N curve with a typical slope of 3. But when the structure is subjected to dynamic loading and the band of periods are closer to the natural period of the structure, then a small change in the assumed period of the applied load significantly changes the result. Marine structures, which are subjected to wind and wave loading, show higher response under certain periods and vice versa. As a result, the number of cycles of stress response may differ from the number of loading cycles, which is handled in the spectral fatigue analysis. Deterministic fatigue analysis is performed using semiempirical relationships. When marine structures are subjected to waves, then only a single wave analysis is used to describe the lifetime stress history of the structure.

5.7.1 Long-Term Exceedance to Be Weibull

Assuming long-term exceedance as a Weibull form, the following relationship holds good:

$$n(\text{stress} > \sigma) = n_{o0} \exp\left\{\left(\frac{\sigma}{\sigma_o}\right)^h \ln(n_{o0})\right\} \tag{5.11}$$

where n is the number of stress cycles exceeding stress in n_o0 cycles, σ_o0 is the stress magnitude that is exceeded once in n_o0 cycles, h is a parameter that depends on the load and response characteristics of the structure (typical value is in the range of 0.5–1).

5.7.2 Long-Term Exceedance to Be Log-Linear

For the long-term exceedance to be log-linear, the following relationship holds good:

$$n(\text{wave height} > H) = n_{o0} \exp\left\{-\left(\frac{H}{H_0}\right) \ln(n_{o0})\right\} \tag{5.12}$$

where H_0 is the wave height exceeded once in the number of known cycles, and H is the wave height exceeded "n" times within the number of known cycles. The long-term exceedance is considered as the S–N curve with a single slope and is used to estimate the fatigue damage in n_L cycles.

For Weibull distribution, damage estimate is given by

$$D_L = \frac{n_L \sigma_o^m}{A}\left[\frac{\Gamma(1+(m/h))}{\ln(n_0)^{m/h}}\right] \tag{5.13}$$

For the wave height exceedance to follow log-linear, damage estimate is given by

$$D_L = \frac{n_L (aH_o^b)^m}{A}\left[\frac{\Gamma(1+bm)}{\ln(n_0)^{bm}}\right] \tag{5.14}$$

where Γ is a gamma function, which is defined as

$$\Gamma(g) = \int_0^\infty x^{g-1} e^{-x} dx \tag{5.15}$$

5.8 SPECTRAL FATIGUE ANALYSIS

Spectral fatigue analysis is applicable to marine structures as they are subjected to dynamic loading that exhibit statistical stationary properties for a large number of stress cycles. The spectral method uses the shape of the stress spectrum to determine the number of stress cycles of various magnitudes and ranges; the chosen stress spectrum can either be a narrow-banded or broad-banded.

5.8.1 Narrow-Banded Spectrum

For a narrow-banded spectrum, zeroth (m_0) and 2nd (m_2) moments of the spectrum, about the line $\{f = 0\}$. It is a usual practice to assume Rayleigh distribution of the stress range in a given stress spectrum. Moments can be computed using numerical integration technique or simple by trapezoidal rule. Zeroth moment (m_0) is the area under the spectrum, which will correspond to the variance of the signal. Square root of the ratio of the moments will give the mean zero-crossing period of the signal, as

$$T_Z = \sqrt{\frac{m_0}{m_2}} \tag{5.16}$$

The number of stress cycles (n) in the time (T) is given by

$$n = \frac{T}{T_z} \tag{5.17}$$

Rayleigh distribution assumes the power spectral density function of the stress range as

$$p(\sigma_r) = \frac{\sigma_r}{4m_0}\exp\left(-\frac{\sigma_r^2}{8m_0}\right) \tag{5.18}$$

In T seconds, which is typically 3 hours sea-state or 1 hour mean-wind duration, the number of stress cycles (δn) in the band ($\delta\sigma_r$), which is centered on σ_r is given by

$$\delta_n = np(\sigma_r)(\delta\sigma_r) \tag{5.19}$$

Fatigue damage associated with that band of stress cycle is given by

$$\delta_D = \frac{\delta_n}{N} = \frac{\delta_n}{A\sigma_r^{-m}} = \frac{n(\sigma_r/4m_0)\exp[-(\sigma_r^2/8m_0)]\delta\sigma_r}{A\sigma_r^{-m}} \tag{5.20}$$

Fatigue damage of all stress cycle bands is found by integrating the above equation as shown below:

$$D = \int_0^\infty \frac{n(\sigma_r/4m_0)\exp[-(\sigma_r^2/8m_0)]\delta\sigma_r}{A\sigma_r^{-m}} \tag{5.21}$$

$$D = \frac{n}{4Am_0} \int_0^{\infty} \sigma_r^{(1+m)} \exp\left(-\frac{\sigma_r^2}{8m_0}\right) \delta\sigma_r \tag{5.22}$$

The integral has a standard solution of S–N curve with a constant slope, which is a gamma function as given by

$$\int_0^{\infty} x^a \exp(-bx^2)dx = \frac{\Gamma((a+1)/c)}{C(B^{((a+1)/c)})} \tag{5.23}$$

The above method gives the fatigue damage for the duration of one particular spectrum. To obtain fatigue damage over a long period, computations need to be repeated for all such spectra that occur within the chosen period. In that case, $\{T\}$ becomes cumulative time for which each spectrum occurs (say, e.g., 1 year). Further, the constant amplitude stress range is replaced by effective fatigue stress range, which is given by

$$\sigma_{efr} = \sqrt{8m_0}\left[\Gamma\left\{\frac{2+m}{m}\right\}\right]^{1/m} \tag{5.24}$$

The effective stress range can also be rewritten using root-mean-square value of the stress range and the significant stress range are given as

$$\sigma_{rms} = 2\sqrt{2m_0} \tag{5.25}$$

$$\sigma_{significant} = \sqrt{4m_0} \tag{5.26}$$

Hence, the effective stress range is given by

$$\sigma_{efr} = \sigma_{rms}\left[\Gamma\left\{\frac{2+m}{m}\right\}\right]^{1/m} \tag{5.27}$$

$$= \sigma_{significant}\frac{1}{\sqrt{2}}\left[\Gamma\left\{\frac{2+m}{m}\right\}\right]^{1/m} \tag{5.28}$$

5.8.2 Broad-Band Spectrum

There are many methods available to explain how to count the stress range cycles in stationary broad-band time history. Rainflow counting accounts for the largest cycles that are extracted first; smaller cycles are considered to be superimposed on the larger cycles. This method is considered to be one of the most reliable methods for fatigue stress range counting. An upper bound method accounts for matching each positive crest with that of the approximate equivalent trough, while the lower bound method matches each crest to that of the following trough. It is important to note that none of the above are useful for frequency domain calculations, as the definition of cycles is set up in terms that are not amenable to statistical analysis.

The spectral fatigue damage analysis of marine structures subjected to random loading assumes that the signal is stationary, Gaussian, and random. Results are generally produced for the mean period of zero crossing per unit time, as given below:

$$T_Z = \sqrt{\frac{m_0}{m_2}} \tag{5.29}$$

For mean time between the peaks or crests per unit time, it is given by

$$T_c = \sqrt{\frac{m_2}{m_4}} \tag{5.30}$$

where m_n is the nth moment of the PSD function and is given by

$$m_n = \int_0^\infty f^n S_{\sigma\sigma}(f)df \tag{5.31}$$

where $S_{\sigma\sigma}(f)$ is the one-sided stress spectrum and f is the frequency in hertz; nth moments are obtained by numerical integration. In predicting the fatigue damage, there is a difficulty in predicting the stress cycle distribution from a known stress spectrum. While this values lies between 0 and 1, for values approaching 1, the signal becomes similar to that of a regular sine wave. At this level, the signal becomes narrow-banded, and the probability density function follows a Rayleigh distribution, which is given by

$$p(\sigma_p) = \frac{\sigma_p}{m_0}\exp\left(-\frac{\sigma_p^2}{2m_0}\right) \tag{5.32}$$

One of the major advantages of the above is that the cycle counting becomes easy and simple. Alternatively, an irregularity factor can be used in the fatigue damage estimates, as given below:

$$\beta = \frac{T_c}{T_z} \tag{5.33}$$

The irregularity factor is important because it accounts for the difficulties of stress cycle distribution from a stress spectrum, which is largely determined by its value lying between 0 and 1. As this factor approaches zero, the signal becomes white noise and is said to be completely wide-banded. The probability density function of the peak follows a Gaussian distribution and is given by

$$p(\sigma_p) = \frac{1}{\sqrt{2\pi m_0}} \exp\left(-\frac{\sigma_p^2}{2m_0}\right) \tag{5.34}$$

In reality, the response is neither narrow- nor completely wide-banded, but lies in between and, therefore, correction factors must be applied to find the solution.

Several researchers have attempted to correct the narrow-band fatigue damage calculation for the effects of broad bandwidths. These corrections are developed by generating sample time histories from stress spectra through inverse Fourier transforms, and then obtaining a conventional rainflow cycle.

5.9 CORRECTION FACTORS FOR FATIGUE DAMAGE ESTIMATES

5.9.1 Wirsching's Correction Factor

The fatigue damage estimate is given by (Wirsching and Light, 1980):

$$D_{RF} = \lambda D_{NR}(m, \in) \tag{5.35}$$

where D_{RF} is the rainflow counted damage and D_{NB} is the estimated damage.

$$\begin{aligned} \lambda(m,\in) &= a(m) + [1 - a(m)](1 - t)^{c(m)} \\ a(m) &= 0.926 - 0.333m \\ c(m) &= 1.587m - 2.323 \\ \in &= \sqrt{1 - \beta^2} \end{aligned} \tag{5.36}$$

5.9.2 Kam and Dover Approach

The idea here is to conceive a complete linearity in the cumulative fatigue damage caused by the constant stress amplitude using rainflow cycles that are extracted from the stress cycle. This expression uses the equivalent stress range parameter, as given by (Kam and Dover, 1989, 1988):

$$\sigma_{efr} = \int_0^\infty \sigma_r^m p(\sigma_r) d\sigma_r \tag{5.37}$$

or

$$\sigma_{efr} = 2\sqrt{2m_0}\left[\lambda(m,\varepsilon)\Gamma\left(\frac{m}{2}+1\right)\right] \tag{5.38}$$

5.9.3 Chaudhary and Dover Approach

Based on the study of peak distribution in different sea state spectra, the following equation is proposed by Chaudhary and Dover (1985):

$$\sigma_{efr} = 2\sqrt{2m_0}\left[\frac{\varepsilon^{m+2}}{2\sqrt{\pi}}\Gamma\left(\frac{m+1}{2}\right)+\frac{\beta}{2}\Gamma\left(\frac{m+2}{2}\right)+err(\beta)\frac{\beta}{2}\Gamma\left(\frac{m+2}{2}\right)\right]^{1/m} \quad (5.39)$$

where

$$\begin{aligned}\text{err}(\beta) &= 0.3012(\beta)+0.4916(\beta)^2+0.918(\beta)^3-2.3534(\beta)^4-3.3307(\beta)^5\\ &\quad +15.654(\beta)^6-10.7846(\beta)^7 \quad \text{for } 0.13<\beta<0.96\end{aligned}$$

5.9.4 Hancock's Equation

Hancock and Gall (1985) proposed an equation to incorporate β and ε into the narrow-band, and this is given by

$$\sigma_{efr} = 2\sqrt{2m_0}\left[\beta\Gamma\left(\frac{m}{2}+1\right)\right]^{1/m} \quad (5.40)$$

$$\sigma_{efr} = \beta\sqrt{2m_0}\,(2-\varepsilon^2)\left[\Gamma\left(\frac{m}{2-\varepsilon^2}+1\right)\right]^{1/m} \quad (5.41)$$

The above factors are used to amend the traditional narrow-band approach. An alternate approach is to avoid the narrow-band assumption and to develop a fatigue life prediction in terms of rainflow ranges. The probability density function of the rainflow ranges is given by

$$P_{RF}(\sigma_r) = \frac{(D_1/Q)e^{-Z/Q}+(D_2Z/R^2)e^{-Z^2/2R^2}+D_3Ze^{-Z^2/2}}{2m_0^{1/2}} \quad (5.42)$$

$$\beta = \frac{T_c}{T_Z} = \sqrt{\frac{m_2^2}{m_0m_4}} \quad (5.43)$$

$$x = \frac{T_c}{T_m} = \frac{m_1}{m_0}\sqrt{\frac{m_2}{m_4}};\ Z = \frac{\sigma_r}{\sqrt{2m_0}} \quad (5.44)$$

$$D_1 = \frac{2(x_m-\beta^2)}{1+\beta^2} \quad (5.45)$$

$$D_2 = \left\{ \frac{1 - \sigma - D_1 + D_1^2}{(1 - R)} \right\} \tag{5.46}$$

$$D_3 = 1 - D_1 - D_2 \tag{5.47}$$

$$Q = \left\{ \frac{1.25(\beta - D_3 - RD_2)}{D_1} \right\} \tag{5.48}$$

$$R = \left\{ \frac{\beta - x_m - D_1^2}{1 - \beta - D_1 + D_1^2} \right\} \tag{5.49}$$

$$m_n = \int_0^\infty f^n S_{\sigma\sigma}(f) df \tag{5.50}$$

To determine the fatigue damage estimate, effective stress range is used as given by Equation 5.37. As suggested by Hancock and Gall (1985), the probability density function is replaced by the one given by Equation 5.42 to obtain the effective stress range. For $n = T/T_c$, damage can be estimated by

$$D = \left(\frac{T}{T_c} \right) \left(\frac{1}{A} \right) \int_0^\infty \sigma_r^m P_{RF}(\sigma_r) d\sigma_r \tag{5.51}$$

5.10 CRACK PROPAGATION

The application of fracture mechanics to the fatigue of steel structures uses the Paris and Erdogan (1983) law. This law states that crack growth (δ_a) in δN cycles in the applied stress range (σ_r) is given by

$$\delta_n = C \left[y \sigma_r \sqrt{\pi a} \right]^m \delta_n \tag{5.52}$$

$$= C(\delta k)^m \delta N \tag{5.53}$$

where y is the crack- and geometry-dependent factor is unity for central, through thickness crack in a wide plate; a is the crack length, which increases with increases in the number of applied stress cycles; and C and m are material-dependent constants. Typical mean values for C and m for BS 4360 grade 50D steel are given below:

$$C = 5.2 \times 10^{-12} \text{ to } 7.1 \times 10^{-12}$$

$$m = 3$$

The unit of C is complex and hence the following table can be used for conversion:

To Convert	From To	Multiply C by
Crack size	m to mm	$10^{3(1-m/2)}$
Stress	MPa to kPa	10^{-3m}
Stress	MPa to Pa	10^{-6m}

Fatigue crack propagation, based on fracture mechanics, is normally worked out in a tabular form. Each column in the table will have values for the set of variables, namely, $\{a, \sigma_r, y, \Delta k, \delta N, \delta a\}$. Each row of the table calculates the crack growth in δN cycles. δN is chosen such that δa is reasonably small in comparison to that of the value of "a." This will result in making the value of "a" as almost constant for the increment of crack growth.

5.10.1 Procedure

- An initial value of crack length "a" is known at the beginning of the computation.
- The stress range may vary for different waves. Hence, use σ_{efr}.

$$\sigma_{efr} = \sqrt[m]{\frac{1}{N}\left(\sum_{i=1}^{N}\sigma_{ri}^{m}\right)}$$

- It is important to note that the growth of the crack leads to a reduction of stiffness and causes redistribution of stresses away from the crack. This requires an additional computation of effective stress for different crack lengths, but is generally ignored.
- Y is calculated at each stage of crack growth.
- $\Delta k = Y\sigma_r\sqrt{\pi a}$.
- δN is selected to give a small change in the crack length. Depending on the rate of crack growth, the ratio of crack growth, δN, may be selected to correspond to the number of cycles in 1 year or 1 month, as the case may be.
- δa is calculated from Equation 5.52.
- "a" is increased from the previous value to "$a + \delta a$."
- Now, $\sigma_r = \sigma_{efr}$, which is the same as in the earlier steps.
- Y is calculated for the new crack size, which is the same as in the earlier steps.
- Hence, $\{a\Delta k, \delta N, \delta a\}$ are computed.
- The calculation is repeated for as many crack growth increments as are required to reach a critical crack size.
- The defect may then be large enough to result in failure if the magnitude of the stress is sufficiently high.

References

Abrahamsen, E. 1976. Safety requirements for offshore engineering. *Behavior of Offshore Structures—First International Conference BOSS '76*, University of Trondheim, Norway, Vol. 1, pp. 877–899.

Adams, A.J. and Baltrop, N.D.P. 1991. *Dynamics of Fixed Marine Structures*. Butterworth-Heinemann Ltd., London.

Anagnostopoulos, S.A. 1982. Dynamic response of offshore structures to extreme waves including fluid-structure interaction. *Engineering Structures*, 4: 179–185.

Ang, A.H.S. and Tang, W.H. 1975a. *Probability Concepts in Engineering and Design, Vol. 1—Basic Concepts*. John Wiley, New York.

Ang, A.H.S. and Tang, W.H. 1975b. *Probability Concepts in Engineering and Design, Vol. 2—Basic Concepts*. John Wiley, New York.

ASTM E 1049-85. 2005. Standard Practices for Cycle Counting in Fatigue Analysis, ASTM International, pp. 1–10.

Bar-Avi, P. 1999. Nonlinear dynamic response of a tension leg platform. *Journal of Offshore Mechanics and Arctic Engineering*, 121: 219–226.

Bea, R.G., Xu, T., Stear, J., and Ramas, R. 1999. Wave forces on decks of offshore platforms. *Journal of Waterway, Port, Coastal and Ocean Engineering*, 125(3): 136–144.

Benjamin, J.R. and Cornell, C.A. 1970. *Probability, Statistics and Decisions for Civil Engineers*. John Wiley, New York.

Bhattacharyya, S.K., Sreekumar, S., and Idichandy, V.G. 2003. Coupled dynamics of Sea Star mini tension leg platform. *Ocean Engineering*, 30: 709–737.

Boaghe, O.M., Billings, S.A., and Stansby, P.K. 1998. Spectral analysis for non-linear wave forces. *Journal of Applied Ocean Research*, 20: 199–212.

Booton, M., Joglekar, N., and Deb, M. 1987. The effect of tether damage on tension leg platform dynamics. *Journal of Offshore Mechanics and Arctic Engineering*, 109: 186–192.

Capanoglu, C.C., Shaver, C.B., Hirayama, H., and Sao, K. 2002. Comparison of model test results and analytical motion analyses for buoyant leg structure. *Proceedings of 12th International Offshore and Polar Engineering Conference*, May 26–31, Kitakushu, Japan, pp. 46–53.

Chandrasekaran, S. 2013a. *Dynamic of Ocean Structures, Web-Based Course on National Program on Technology Enhanced Learning*. IIT Madras, nptel.ac.in/courses/114106036

Chandrasekaran, S. 2013b. *Ocean Structures and Materials, Web-Based Course on National Program on Technology Enhanced Learning*. IIT Madras, nptel.ac.in/courses/114106035

Chandrasekaran, S. 2013c. *Advanced Marine Structures, Web-Based Course on National Program on Technology Enhanced Learning*. IIT Madras, nptel.ac.in/courses/114106037

Chandrasekaran, S. and Bhattacharyya, S.K. 2012. *Analysis and Design of Offshore Structures with Illustrated Examples. Human Resource Development Center for Offshore and Plant Engineering (HOPE Center)*. Changwon National University Press, South Korea, ISBN: 978-89-963915-5-5.

Chandrasekaran, S., Bhaskar, K., and Hashim, M. 2010b. Experimental study on dynamic response behavior of multi-legged articulated tower. *Proceedings of the 29th International Conference on Ocean, Offshore and Arctic Engineering, OMAE 2010*, June 6–11, Shanghai, China, pp. 1–10.

Chandrasekaran, S., Bhaskar, K., Harilal, L., and Brijit, R. 2010a. Dynamic response behavior of multi-legged articulated tower with and without TMD, *Proceedings of the International Conference of Marine Technology. MARTEC-2010*, December 11–12, Dhaka, Bangladesh, pp. 131–136.

Chandrasekaran, S., Chandak, N.R., and Gupta, A. 2006a. Stability analysis of TLP tethers. *Ocean Engineering,* 33(3): 471–482.

Chandrasekaran, S. and Gaurav. 2008. Offshore triangular tension leg platform earthquake motion analysis under distinctly high sea waves. *Journal of Ships and Offshore Structures,* 3(3): 173–184.

Chandrasekaran, S. and Jain, A.K. 2002a. Dynamic behavior of square and triangular TLPs under regular wave loads. *Ocean Engineering*, 29(3): 279–315.

Chandrasekaran, S. and Jain, A.K. 2002b. Triangular configuration tension leg platform behaviour under random sea wave loads. *Ocean Engineering*, 29(13): 1895–1928.

Chandrasekaran, S. and Jain, A.K. 2004. Aerodynamic behavior of offshore triangular tension leg platforms. *Proceedings of ISOPE*, Toulon, France, pp. 564–569.

Chandrasekaran, S., Jain, A.K., and Chandak, N.R. 2004. Influence of hydrodynamic coefficients in the response behavior of triangular TLPs in regular waves. *Ocean Engineering*, 31: 2319–2342.

Chandrasekaran, S., Jain, A.K., and Chandak, N.R. 2006b. Seismic analysis of offshore triangular tension leg platforms. *International Journal of Structural Stability Dynamics*, 6(1): 97–120.

Chandrasekaran, S., Jain, A.K., and Chandak, N.R. 2007a. Response behavior of triangular tension leg platforms under regular waves using Stokes nonlinear wave theory. *Journal of Waterway Port Coastal Ocean Engineering ASCE*, 133(3): 230–237.

Chandrasekaran, S., Jain, A.K., Gupta, A., and Srivastava, A. 2007b. Response behavior of triangular tension leg platforms under impact loading. *Ocean Engineering*, 34: 45–53.

Chandrasekaran, S. and Pannerselvam, R. 2009. Offshore structures: Materials, analysis, design and construction, *Proceedings of Short Course on Offshore Structures and Materials*, December 14–18, IITM, Chennai, pp. 156.

Chandrasekaran, S. and Saha, N. 2011. Reliability of offshore structures, Short term course. Department of Ocean Engineering, IIT Madras, India.

Chandrasekaran, S., Sharma, A., and Srivastava, S. 2007c. Offshore triangular TLP behavior using dynamic Morison equation. *Journal of Structural Engineering,* 34(4): 291–296.

Chandrasekaran, S., Sundaravadivelu, R., Pannerselvam, R., and Madhuri, S. 2011. Experimental investigations of offshore triceratops under regular waves, *Proceedings of the 30th International Conference on Ocean, Offshore and Arctic Engineering, OMAE 2011*, June 19–24, Rotterdam, the Netherlands.

Chakrabarti, S.K. 1987. *Hydrodynamics of Offshore Structures: Computational Mechanics.* p. 440, ISBN: 9780387173191.

Chakrabarti, S.K. 1990. *Non-Linear Method in Offshore Engineering.* Elsevier Science, the Netherlands.

Chakrabarti, S.K. 1994. *Offshore Structure Modeling, Advanced Series on Ocean Engineering*, 9, World Scientific, Singapore.

Chaudhary, G.K. and Dover, W.D. 1985. Fatigue analysis of offshore platforms subjected to sea wave loading. *Proceedings of the International Journal of Fatigue*, 7(1): 13–19.

Choi, H.S. and Lou, J.Y.K. 1991. Nonlinear behavior of an articulated offshore loading platform. *Applied Ocean Research*, 13(2): 63–74.

Clauss, G.F. and Birk, L. 1996. Hydrodynamic shape optimization of large offshore structures. *Journal of Applied Ocean Research*, 18: 157–171.

Clauss, G.T. et al. 1992. *Offshore Structures, Vol. 1—Conceptual Design and Hydromechanics*, Springer, London.

Cornell, C.A. and Vanmarche, E.H. 1975. Seismic risk analysis for offshore structures. *Offshore Technology Conference*, Houston, Paper OTC 2350.

Daniels, H.E. 1945. The statistical theory of the strength of bundles of threads. *Royal Statistical Society Series A,* 183: 405.

Dawson, T.H. 1983. *Offshore Structural Engineering*, Prentice-Hall Inc, New Jersy.

de Boom, W.C., Pinkster Jo. A, and Tan, P.S.G. 1984. Motion and tether force prediction of a TLP. *Journal of Waterway, Port, Coastal and Ocean Engineering*, 110(4): 472–486.

Demirbilek, Z. 1990. Design formulae for offset, set down and tether loads of a tension leg platform (TLP). *Ocean Engineering*, 17(5): 517–523.

Ditlevsen, O. 1977. *Fundamentals of Second Moment Structural Reliability Theory. International Research Seminar on Safety of Structures under Dynamic Loading*, Trondheim, Norway.

Ertas, A. and Eskwaro-Osire, S. 1991. Effects of damping and wave parameters on offshore structure under random excitation. *Nonlinear Dynamics*, 2: 119–136.

Fjeld, S. 1977. Reliability of offshore structures. *Offshore Technology Conference*, Houston, Paper OTC 3027.

Freudenthal, A.M. and Gaither, W.S. 1969. Design criteria for fixed offshore structures. *Offshore Technology Conference*, Houston, Paper OTC 1058.

Furnes, O. 1977. Some aspects of reliability of offshore structures, the need for data. *2nd International Conference on Structural Safety and Reliability*, Munich, Germany, pp. 363–366.

Gerwick, B.C. Jr. 1986. *Construction of Offshore Structures*, John Wiley, New York.

Graff, W.J. 1981. *Introduction to Offshore Structures: Design, Fabrication and Installation*, Gulf Publishing Co, Tokyo.

GustoMSC. 2010. The exploration market. Trends in deepwater drilling. *InSide*, 15: 4–7.

Haldar, A. and Mahadevan, S. 2000. *Probability, Reliability and Statistical Methods in Engineering Design*, John Wiley and Sons, New York.

Halkyard, J.E., Davies, R.L., and Glanville, R.S. 1991. The tension buoyant tower: A design for deep water. *23rd Annual Offshore Technology Conference*, May 6–9, Houston, Texas, pp. 41–55.

Hancock, J.W. and Gall, D.S. 1985. Fatigue under narrow and broad band stationary loading, Final report of cohesive program of R&D into Fatigue of offshore structures. Marine Technology Directorate Ltd, Texas, USA.

Haritos, N. 1985. Modeling the response of tension leg platforms to the effects of wind using simulated traces. *Mathematics and Computers in Simulation*, 27: 231–240.

Hitchings, G.A., Bradshaw, H., and Labiosa, T.D. 1976. Planning and execution of offshore site investigations for North Sea gravity platform. *Proceedings of Offshore Technology Conference*, Paper No. 2430.

Høeg, K. and Tong, W.H. 1977. Probabilistic considerations in the foundation engineering for offshore structures. *2nd International Conference on Structural Safety and Reliability*, Munich, Germany, pp. 267–296.

Hsu, H.T. 1981. *Applied Offshore Structural Engineering*, Gulf Publishing Co., Houston.

Islam, N. and Ahmad, S. 2003. Nonlinear seismic response of articulated offshore tower. *Journal of Defence Science*, 53(1): 105–113.

Issacson, M. and Det St, Q. 1982. Non-linear wave effects on fixed and floating bodies. *Journal of Fluid Mechanics*, 120: 267–281.

Iwaski, H. 1981. *Preliminary Design Study of Tension Leg Platform*, MIT University, USA.

Kam, J.C.P. and Dover, W.D. 1988. Fast fatigue assessment procedure for offshore structure under random time history. *Proceedings Institution of Civil Engineers*, 85(Pt 2): 689–700.

Kam, J.C.P. and Dover, W.D. 1989. Advanced tool for fast assessment of fatigue under offshore random wave stress history. *Proceedings Institution of Civil Engineers*, 87(Pt 2): 539–556.

Kobayashi, M., Shimada, K., and Fujihira, T. 1987. Study on dynamic responses of a TLP in waves. *Journal of Offshore Mechanics and Arctic Engineering*, 109: 61–66.

Love, A.E.H. 1994. *Mathematical Theory of Elasticity*, Dover Publications Inc, New York.

Low, Y.M. 2009. Frequency domain analysis of a tension leg platform with statistical linearization of the tendon restoring forces. *Marine Structures*, 22: 480–503.

Madsen, H.O., Krenk, S., and Lind, N.C. 2006. *Methods of Structural Safety*, Dover Publications Inc, New York.

Marshall, P.W. 1969. Risk evaluations for offshore structures. *Journal of the Structural Division ACSE*, 95: 2907–2929.

Marshall, P.W. 1976. Failure modes for offshore platforms—fatigue. *Behavior of Offshore Structures—First International Conference BOSS '76*, University of Trondheim, Norway, Vol. 2, pp. 234–248.

Marshall, P.W. and Bea, R.G. 1976. Failure modes of offshore platforms. *Behavior of Offshore Structures—First International Conference BOSS '76*, University of Trondheim, Norway, Vol. 2, pp. 579–635.

Mather, A. 2000. *Offshore Engineering: An Introduction*, 2nd ed., Witherby & Co., Ltd, London, U.K.

Matsuishi, M. and Endo, T. 1968. Fatigue of metals subjected to varying stresses. *Proceedings of Japan Society of Mechanical Engineers,* Fukuoka, Japan, 3: 37–40.

Melchers, R.E. 1999. *Structural Reliability: Analysis and Prediction*, 2nd ed., John Wiley, West Sussex, London, p. 437.

Meyerhof, G.G. 1976. Concepts of safety in foundation engineering ashore and offshore. *Behavior of Offshore Structures—First International Conference BOSS '76*, University of Trondheim, Norway, Vol. 1, pp. 900–911.

Moan, T. and Sigbjørnson, R. 1977. Stochastic sea load effect analysis for probabilistic design of fixed offshore platforms. *2nd International Conference on Structural Safety and Reliability*, Munich, Germany, pp. 227–246.

Moe, G. and Verley, R.L.P. 1980. Hydrodynamic damping of offshore structures in wave and currents. *Offshore Technology Conference, 12th Annual OTC*, Houston, Texas, pp. 37–44.

Moses, F. 1976. Reliability of structural systems. *Behavior of Offshore Structures—First International Conference BOSS '76*, University of Trondheim, Norway, Vol. 1, pp. 912–923.

Moses, F. 1977. Safety and reliability of offshore structures. *International Research Seminar on Safety of Structures under Dynamic Loading*. Trondheim, Norway.

Moses, F. and Stevenson, J. 1970. Reliability-based structural design. *Journal of the Structural Division ACSE*, 96: 221–244.

Naess, A. and Moan, T. 2013. *Stochastic Dynamics of Marine Structures*, Cambridge University Press, New York, USA.

Nagamani, K. and Ganapathy, C. 2000. Dynamic response of three leg articulated tower. *Ocean Engineering*, 27: 1455–1471.

Neville, A.M. 1997. *Properties of Concrete*, 4th ed., John Wiley & Sons, New York.

Nordgren, R.P. 1987. Analysis of high frequency vibration of tension leg platforms. *Journal of Offshore Mechanics and Arctic Engineering*, 109: 119–125.

Nordic Committee for Building Regulations (NKB). 1977. *Proposal for Safety Codes*. TS42, September.

Paik, I. and Roesset, J.M. 1996. Use of quadratic transfer functions to predict response of tension leg platforms. *Journal of Engineering Mechanics*, 122(9): 882–889.

Paik, J.K. and Thayamballi, A.K. 2007. *Ship-Shaped Offshore Installations: Design, Building and Operations*. Cambridge University Press, New York.

Papoulis, A. and Pillai, S.U. 1991. *Probability, Random Variables and Stochastic Processes*, 3rd ed., McGraw-Hill, New York.

Patel, M.H. 1989. *Dynamics of Offshore Structures.* Butterworth, London.

Pavia, A.P., Yin, P.K., and Corder, P.R. 1977. Effect of design, fabrication, and installation on the structural reliability of offshore platforms. *Offshore Technology Conference*, Houston, Paper OTC 3026.

Perryman, S.R., Horton, E.E., and Halkyard, J.E. 1995. Tension buoyant tower for small fields in deep waters. *Offshore Technology Conference*, May 1–4, Houston, Texas, pp. 13–22.

Popovics, J.S., Zemajtis, J., and Shkolnik, I. 2008. Studies on static and dynamic modulus of elasticity, ACI-CRC report.

Rackwitz, R. 1977. First order reliability theories and stochastic models. *2nd International Conference on Structural Safety and Reliability*, Munich, Germany, pp. 199–220.

Reddy, D.V. and Arockiasamy, M. 1991. *Offshore Structures, Vol. I.* Kriger Publishing Co., Malabar, Florida.

Rijken, O.R., Niedzwecki J.M. 1991. *A Knowledge Base Approach to the Design of Tension Leg Platform.* Offshore Technology Center, Texas, pp. 24–100.

Robert, W.C. and Capanoglu, C.C. 1995. Buoyant leg structure for the development of marginal fields in deep water. *Proceedings of International Offshore and Polar Engineering Conference*, June 11–16, The Hague, The Netherlands, pp. 163.

Røren, E.M.Q. and Furnes O. 1976. State of the art: Behaviour of structures and structural design. *Behavior of Offshore Structures—First International Conference BOSS '76*, University of Trondheim, Norway, Vol. 1, pp. 70–111.

Røren, E.M.Q. and Furnes, O. 1977. Performance of offshore structures—State of the art. *Norwegian Maritime Res*, 5: 14–29.

Sadehi, K. 1989. *Design and Analysis of Marine Structures.* Khajeh Nasirroddin Tsi University of Technology, Tehran, Iran.

Sadehi, K. 2001. *Coasts, Ports and Offshore Structures, Engineering.* Power and Water University of Technology, Tehran, Iran.

Sadehi, K. 2007. Offshore and petroleum platforms for Cyprus oil/gas fields. *GAU Journal of Social & Applied Sciences*, 2(4): 1–16.

Sarpkaya, T. and Isaacson, M. 1981. *Mechanics of Wave Forces on Offshore Structures.* Van Nostrand Reinhold.

Schuêller, G.I. and Choi, H.S. 1977. Offshore platform risk based on a reliability function model. *Offshore Technology Conference*, Houston, Paper OTC 3028.

Shaver, C.B., Capanoglu, C.C., and Serrahn, C.S. 2001. Buoyant leg structure preliminary design, constructed cost and model test results. *Proceedings of 11th International Offshore and Polar Engineering Conference*, June 17–22, Stavanger, Norway, pp. 432–439.

Subramaniam, K.V., Popovics, J.S., and Shah, S.P. 2000. Determining elastic properties of concrete using vibration resonance frequencies of standard test cylinders. *Cement Concrete Aggregates (ASTM)*, 22(2): 81–89.

Tabeshpour, M.R., Golafshani, A.A., and Seif, M.S. 2006. Comprehensive study on the results of tension leg platform responses in random sea. *Journal of Zhejiang University Science*, 7(8): 1305–1317.

Thoft-Christensen, P. 1977. Some experience from application of optimization technique in structural reliability. *Bygningsstatiske Meddelelser*, 48: 31–44.

Throft-Christensen, P. and Baker, M. 1982. *Structural Reliability Theory and Applications.* Springer Verlag, Berlin.

Vannucci, P. 1996. Simplified optimal design of tension leg platform TLP. *Structural Optimization*, 12: 265–268.

White, N.C., Copple, W.R., and Capanoglu, C. 2005. Triceratops: An effective platform for developing oil and gas fields in deep and ultra-deep water. *Proceedings of 15th International Offshore and Polar Engineering Conference*, June 19–24, Seoul, Korea, pp. 133–139.

Wirsching, P. and Ortiz, P.K. 2006. *Random Vibration: Theory and Practice*, Dover Publications Inc, New York.

Wirsching, P.H. and Light, M.C. 1980. Fatigue under wide band random loading. *Journal of the Structural Design, ASCE*, 106(ST7): 1593–1607.

Witz, J.A., Patel, M.H., and Harrison, J.H. 1986. On the hydrodynamics of semisubmersibles with articulated members. *Proceedings of Royal Society of London, Series A, Mathematical and Physical Sciences*, 403: 81–109.

Yan, F.-S., Zhang, D.-G., Sun, L.-P., and Dai, Y.-S. 2009. Stress verification of a TLP under extreme wave environment. *Journal of Marine Science Applications*, 8: 132–136.

Yang, J.N. and Freudenthal, A.M. 1977. Reliability assessment of offshore platforms in seismic regions. *2nd International Conference on Structural Safety and Reliability*, Munich, Germany, pp. 247–266.

Yoneya, T. and Yoshida, K. 1982. The dynamics of tension leg platforms in waves. *Journal of Energy Resources Technology*, 104: 20–28.

Yoshida, K., Ozaki, M., and Oka, N. 1984. Structural response analysis of tension leg platforms. *Journal of Energy Resources*, 106: 10–17.

Young, A.G., Kraft, L.M., and Focht, J.A. 1975. Geotechnical considerations in foundation design of offshore gravity structures. *Proceedings of Offshore Technology Conference*, Paper No. 2371.

Index

G

H

I

J

P

R

Y